TIME AND THE APPLICATION OF TIME

Samuel K. K. Blankson

Published by Blankson Enterprises Ltd., www.practicalbooks.org. This book is made available through Print on Demand.

A copy of this title is held by the British Library.

Paperback Edition (9 x 6)
ISBN 13: 978-1-4457-2501-7

Also available in e-book and hardback format from Lulu.com.

Other works by Samuel K. K. Blankson

The metaphysical foundations for physics
Why time is not a natural phenomenon
The mathematical theory of time
How old is the universe in time
The Einstein theory of space-time
Without mathematics
Dead end (a novel)
Time in science and life
The greatest legacy of Albert Einstein
How religious scientists play down the greatest of Einstein's achievements
The coming revolution in physics

Contents

PROLOGUE

This little book is yet another of my efforts seeking to clarify the nature of time and its uses. As is obvious, I write in very small volumes due to financial constraints. I also write specifically for the general reader. The deliberately cultivated (Russellian) simple style I use could bore some professionals and specialists to death. They rather prefer the more technical manner of writing, where things would get so difficult that the general reader could hardly understand a thing. Nevertheless I quite agree with them; it makes life easy for the specialists. On the contrary, this mode of presentation is rather difficult for the writer and easy for the reader---that is the irony of the matter. But I take the view that, although time is very technical in science, and much more difficult to write about in plain language, but it is known and used by everybody. Therefore I want to write in a language accessible to every general reader and thereby create a quandary: the specialists may be put off as they tend to hate lucidity in technical subjects; yet readers like you and I want to be able to follow what is discussed about time because it affects all of us equally, and therefore the most important subject in all life. Only issues concerning life itself can take precedence over discussions of time; and those who believe that mathematics is difficult should try to ponder the nature of time by the laws of logic.

In the end, it may very well be a penalty for me if the specialists who could confirm my suppositions are put off, to deny me the attention and wealth and prime-time TV interviews that I may deserve; in a word, the current distasteful fetish with success even in intellectual matters of this abstract nature. I am not unaware that intellectually this is the most serious academic (philosophical, scientific) subject in the world but, at my age, being in my seventies, I need nothing and want nothing---just the truth. I was shocked when one journal editor rejected my paper not because my theory was untrue or wrong, but, as he put it, "you rely on semipopular books as your sources." Those "unacceptable" semipopular books included Einstein's own book "Relativity", and Bertrand Russell's "Analysis of Matter", Professor Eddington's "Mathematical Theory of Relativity", etc. In the process the truth was sacrificed in the name of insular and clearly obsolescent editorial policy.

This evokes what I call "The mathematical paradox of scientific discovery in physical theory": the inventors of mathematical techniques usually have no problem in their sight. To them it is a playful game or pastime. Those who apply them to problems must perforce pick and choose the suitable mathematics to apply. If none is available, progress is stalled; yet what is consistent with existing mathematics (not to be found in 'semipopular books') may not be true of the world and will eventually bring down part of our scientific edifice—e.g. the eather debacle. So is it right to reject what is published in 'semipopular books' like "Relativity" by Einstein?

Be that as it may, in this monograph, as far as the general reader is concerned, actually the title 'Time and the Application of Time', generally, as it might be supposed, is really a misnomer. Rather it is in the fields of theoretical physics, taken as including astronomy and cosmology (and also in technical philosophy for the interpretation of physical reality), that the nature of time as opposed to its specific applications requires clarifications.

To the ordinary man, like you and me, time is simply time in the clock; but the philosophers know that it came from somewhere to get there. Clocks are manufactured by human beings; clocks were not in existence when life began. And whoever created the world or the so-called intelligent design failed to provide us with time already in a clock. Thus Bertrand Russell asked the most important question about life and time: if, by the Einstein theory, cosmic time is abandoned, then what is measured by the clock? I give my answer in Appendix I.

However, we know that time is quite distinct from how it is employed: for cooking, for earning money from work, for attending church, hospital appointments, going to school, colleges and universities, and so forth. We know these as the applications of time: a physical entity in existence that we use for various purposes---distinct from those purposes, none of which can be said to reveal the true nature (or metaphysical nature) of the time. Normally we do not make the mistake of confusing the time with its uses—that is, conflating its existence with its applications. Time is one thing; its application is quite another. It is my belief that if you learn to separate time as it is (provided you can define it logically as I will show in this book), from its applications, or uses we put it to, you will find that most of the mysteries about time disappear. Scientists have overlooked the difference between time, as an entity, and its applications as another distinct operation of the human mind; but, ironically, all mankind know of the applications of time only, as they also err on the ignorance of the existence of time as a distinct entity in nature. For this reason people entertain many legends about time that knowledge of its nature as a distinct entity in nature would make them realise that they are mistaken---but who cares? People just use time.

On the other hand, mathematicians claim to have found that time is the same thing as space. And they claim to have come by this knowledge through using time to analyse Order and Simultaneity.[1] By some of the

---

[1] Einstein used this method but only to show that time is rather dynamic and not absolute in the Newtonian sense, not that it is one and the same thing as space. He could not have

very advanced mathematical techniques known to man, they assert that time is seen as one and the same thing as any space. The summary of the mathematical equations (dozens of them) which I will write as '*s=ct...*' throughout this monograph, is meant to say s is for space, and it is equal to time, 'ct', as seen by means of light. This short form of the very complicated equations will do, as it gives the gist of what we want to discuss, namely, is time independent of space or it is the same thing as space in a world of 4-D geometry? The following quotation may help: "Having already rocked the mathematical world to its foundations with his incompleteness theorem, Gödel now took aim at Einstein and relativity. Wasting no time, he announced in short order his discovery of new and unsuspected cosmological solutions to the field equations of general relativity, solutions in which time would undergo a shocking transformation. The mathematics, the physics, and the philosophy of Gödel's results were all new. In the possible worlds governed by these new cosmological solutions, the so-called rotating or Gödel universes, it turned out that the space-time structure is so greatly warped or curved by the distribution of matter that there exist timelike future-directed paths by which a spaceship, if it travels fast enough...can penetrate into

---

imagined that because he was researching time in the knowledge that Lorentz had already discovered that time can begin from anywhere and so the Newtonian notion of time was wrong. He was not concerned with the origin of time---that is either from space or wherever. Time was time in the clock, and by using it to analyse Order and Simultaneity, he found that it was changeable and not absolute. That is very different from what scientists subsequently used the same method to do, namely to show that time is the same as space in a world of 4-D geometry---based on the imaginary time coordinates of Hermann Minkowski in his *ict* equation.

any region of the past, present or future."[2] Yet if the logical foundation of the Minkowski four-dimensional space is fictitious, then all this is nonsense, and I hereby declare that that is what it is---sheer nonsense about time travel based on flawed logic.

Here is a footnote about the Minkowski theory that must be emphasised to avoid confusion and error. Minkowski did not actually arrive at his supposition through the analysis of Order and Simultaneity; only Einstein did so. However, Minkowski took over the time system discovered by Einstein (and Lorentz) through the analysis of Order and Simultaneity---which is the revolutionary notion that time can begin from anywhere and therefore it is neither general nor absolute. Until then nobody ever thought about questioning (seriously in logic) the Newtonian notions of general and absolute time. In other words, the Einstein analysis was the essential catalyst that introduced the new idea (or definition) of time, which led Minkowski to his theory of 4-D geometry. That is why Russell has noted that the space-time idea was already implied in special relativity. I think he meant it in the sense that time which can be started from anywhere obviously owes something to space---but that it comes to exist only as a union between time (again) and space is the sore point hotly disputed in this book. In a nutshell, time (in hand already) cannot be combined with space (by means of mathematics) to create another time as 'space-time', since there can be only one time overall. This is a matter of logic, not mathematics; and logically the Minkowski formula is flawed. Since logic is the only intellectual tool with which to carry out mathematical deductions, his complex mathematical edifice is sheer humbug, accepted *only* as a matter of faith by scientists. I know

---

2 From, "A WORLD WITHOUT TIME..." By Professor Palle Yourgrau, Penguin Books, 2007---Ch.1.

that scientists, led by the Royal Society, who adore the Minkowski proposal, will not love me for pointing this out---but they cannot stop me, and so I will go on and on and on. Alas, I don't need their honours. This must be extremely embarrassing for the scientific establishment whose heirs to the great think they are automatically great as well. Given the phenomenal and general depth by which Minkowski dominates physics, astronomy and cosmology, I think they want somewhere to hide when they read me. The question, put simply, is this: is space four-dimensional (including time as we know it?) The answer is, logically, no. So a whole lot of scientific theories needs revising. But the matter is so serious that nobody wants to look at it, let alone scrutinise the case against 4-D geometry by which time travel is said to be 'a scientific possibility---as predicted by special relativity.' The religious instinct in man is so strong that scientists force themselves to believe this because it seems to imply that certain mysteries are beyond scientific scrutiny---therefore the question of God cannot be entirely ruled out, hurray!

I have already published a number of small books to try and refute this proposition without mathematics, and the high and mighty in science have done their best to ignore me, mostly because I have failed to refute the notion with mathematics, or show that any of the equations are faulty. In my own defence, I point out that I do not fail to show that any of the equations are faulty, since I have argued that the basic Minkowski equation is known as 'ict...' (which, to keep out of trouble, I deliberately took care to copy from the master himself, Albert Einstein), and it is faulty because it is supposed to represent time, yet i in the most advanced mathematics known to man refers to imaginary quantities or qualities, whereas time is in no way imaginary. In other words, I am using logic to undermine the equation upon which the whole Minkowski edifice is based—but no dice! Yet it is a mistake to ignore logic in the matter of the origin, essence and passage of time; all of which I think can now be completely resolved by considering time as logical rather than mathematical. I believe Minkowski was wholly mistaken to consider time geometrical---even then only with the help of imaginary time coordinates, an insult to philosophers. Philosophy, as defective as some scientist think it is, is nevertheless required for the rational regulation of science; science cannot progress without philosophical guidance. The reason is logic. And here I must make plain that logic is the ultimate arbiter of truth in the study of all aspects of time, except in its display or tabulations in clocks. Now, we all agree that mathematics, or arithmetic to be exact, plays a part in the process of having time, not necessarily in its applications. But, at the same time, it must be realised that time cannot be wholly mathematical. Minkowski claim to have made it geometrical; but then what about the internal sense of time we know as the sense of duration? In fact, duration is the seat of time "during the life of an event" means time is going. To know how much time, as I will explain later on in this book, we have to use some mathematics because we can only know that by the use of repetitive or regular cycles---and count them, like the earth's repetitive orbits of the sun, as years, and so forth. As I will show later on, that is all we can ever know of time. Nobody can tell what time is---all we can ever do is to show how much of it is passing: the passing of the years is the passing of time building up all the way to the centuries. However, the year is only a physical activity. It is the physical orbit of the sun that we call 'a year', and count them as the passing of time---all we can know of time.

Mathematics is important, but it is from logic that we can structure our time system to accord with the essential features of the world---like the day and night system for instance. The whole of our earth time is organised in such a manner that we know what time is relevant to day or night, as they have cultural implications: you would not normally go out into the bush when it is midnight. If you do, you will stand a good chance of meeting the undertaker! Actually, I believe that time anywhere in the universe will resemble time on earth because of logic. By logic I am referring to rational thought consistent with the features of the world. For the time will have to accord with the features of physical reality or there will be doom. Since rational thought dictates the avoidance of danger, all time systems will be designed to accord with physical reality. This means, since every planetary system will have a sun or parent star, some sort of day and night system will be in operation. And that means any time system in the universe will resemble time on earth. Man is not as stupid as some mathematicians tend to make him out.

Also, I have argued at length that even Bertrand Russell called the Minkowski proposal arbitrary for the same reason, namely that he bases his theory on imaginary time coordinates. Yet still the answer is *nem*, no dice! Philosophical opinions are not well regarded in scientific circles. One gets the impressions that Mathematicians in particular are not normal people; they are all of them half-sane and half-mad, but very, very clever, working entirely through the imagination like dreamers because insanity is pretty close to genius. We rein them in only by logic, and logic alone, and that is what I am trying to do to no effect. Sometimes mathematicians write to remind me of the ds^2 formula which is supposed to include time in its space because of the Minkowski theory; or they point out that Minkowski said this or that, which I always find elementary, patronising and annoying. But the result is zilch. Only one writer from Holland wrote to tell me recently that he agrees with me, and the Polish, too, are writing an interpretation of my work---but from Britain, *nein*. Even my letters to the Royal Society are never acknowledged—OK, *danke!* I can joke about it because they rather inspire me to want to continue, happy in the knowledge that my very existence irritates them.

I mentioned 'very advanced mathematical techniques known to man' above. Well, it is true, but there is a problem; for I have also argued that mathematics, or just because something is mathematical, does not mean it is true of physical reality. This is because mathematicians like to dream and persuade us that it is their own true representation of what reality is---what is one to do with all those brilliant, beautiful and complex (post-professorial) theses in mathematics about the eather for the propagation of light? They are all there preserved in the archives of the great journals, yet they are, or were, useless.

I think mathematics must not be used, or relied upon absolutely in the determination of the nature of physical reality---for the simply reason that these half-mad people cannot be trusted absolutely! Quite seriously, mathematics cannot be used to alter the fundamental entities in nature without physical divisions or practical unions. In Appendix II (culled from other works and attached to this book as an act of permissible literary larceny), I have given my theory of mathematics, called 'The Principle of Mathematical Equivalence' to the effect that mathematics can only mirror reality not change it materially. It can represent reality as it is, but can never alter it physically (that is to say, we cannot do so with mathematics alone).

We know, of course, that by some of those advanced mathematics known to man actually only began with Hermann Minkowski (specifically about time and space), and which were hailed as brilliant, and even praised by Einstein after his initial objections),[3] time and space are

---

[3] He wrote in his seminal book, "RELATIVITY": "It must be clear even to the non-mathematician that, as a consequence of

said to be merged into one entity, or revealed to be one and the same thing by means of mathematics alone in a world of 4-D geometry, so that the whole of theoretical physics, astronomy and cosmology start from the premise that '*s=ct...*' Thus all time is space time, and all space is also space time, leading to some pretty bizarre conclusions, the most shocking (and most beloved) of which is 'curved space time', by which time travel is regarded as really feasible by our most serious scientific thinkers and writers.[4]

The whole of theoretical physics, astronomy and cosmology start from the premise that '*s=ct...*' They work on the assumption that all time is automatically the same thing as space---because Minkowski said *s=ct*. And this is said to have been discovered through the analysis of Order and Simultaneity. However, since they

---

this purely formal addition to our knowledge, the theory [of relativity] perforce gained clearness in no mean measure. These inadequate remarks can give the reader only vague notion of the important idea contributed by Minkowski. Without it the general theory of relativity, of which the fundamental ideas are developed in the following pages, would perhaps have got no farther than its long clothes". He adds, "Minkowski's work is doubtless difficult of access to anyone inexperienced in mathematics..." *Danke*! (See page 58 of his book, reissued in the Routledge Classics, 1993). Yet, originally, he described the mathematical interpretation of his work as 'superfluous learnedness'. In fact, it was worse than that. It was superfluous distortion of the sort of physical reality that can be inferred from the theory of relativity.

[4] Of course, one must agree that if 'curved space time' is the real physical nature of reality, then time can curve to no end so much so that going backwards in centuries, or forwards in thousands of years, can be imagined, even if not, as yet, practically achievable. But, on the other hand, if it is true that that supposition from Minkowski is fictitious, as Professor Eddington called it, then the whole debate is exercise in futility and sheer humbug, or religious nonsense.

have to use time to do that analysis, even if true, the merging of time and space is on a shallow level of reality and of no philosophical significance whatsoever. To say that it makes reality 4-Dimensional (including time) is to overlook the fact that you had the time already before you did the analysis of Order and Simultaneity; so the analysis cannot show that the time you had in hand before your analysis is in fact part of space by means of your analysis. Either this is nonsensical or shallow and of no significance.

Time and space cannot be the only means by which man can get his time, as Minkowski prescribed, for, after all, *mathematics cannot be used to alter physical reality, only to mirror or represent it "as it is" in symbols as a kind of special language. That is all we use mathematics for.* Yet Minkowski specifically said only a union of the two (space and time) can preserved their existence in a new format as "space-time". This means time comes into existence only after the union of time and space into "space-time". How nobody has noticed that this is the worst kind of school boy tautology beats my understanding.

My basic argument in this little book is that scientists have over-looked the fact that there is time, and there are the many applications of the time---two completely different operations of the human mind; but they have always been conflated together, leading to some bizarre suppositions and conclusions about time, the worst of which is the Minkowski concept of space-time as the only way to get time.

This booklet includes five Appendices already published with my previous monographs. I don't want to rewrite them; and yet I believe the reader needs their thought expansions for the proper understanding of my theory of time.

However, for the length of the Introduction for such a small book, I have to apologise with the explanation that the subject is so difficult that I want to introduce it properly with as much details as the reader may require for an effortless understanding in so far as possible---*danke*!

As always, my thanks go to my very clever and compassionate son, Samuel Blankson, the computer buff and prolific writer (with over twenty titles to his name), who acts for me as my agent and publisher at the same time. Unfortunately for him, he had the temerity to complain to a reader in Holland that I do not present my ideas in the traditional format, and was put down with the retort that, that may be so "but it is bearable because the essence is revealing". I rest my case---with this, clearly justifiable, display of occupational vanity!

Samuel K. K. Blankson

Accra and London

CHAPTER I:

THE COMING REVOLUTION IN PHYSICS

The article below is reprinted from a Newspaper without redaction. It was originally written on the advice of my agents in Ghana for the purpose of promoting two little booklets about time based on the Minkowski theory but they were later scrapped as worthless both in content and production. I have since published another booklet about the topic, 'The Coming Revolution in Physics', but failed to reproduce the original articles upon which subject the whole book was based.

I have now realised that, having published it, moreover in a national Newspaper, the contents must be made available to future scholars who might wish to comment on my work as a whole; for the gist of the article was that physics is heading for another revolution due to the misunderstanding and misuse of time because of the concept of 4-D geometry, and therefore the Minkowski theory must be discarded. Looking back, I am pretty certain that many of the mathematicians who saw it probably laughed at me as a lunatic. Also, as published in the paper, it was practically ruined with misprints. In particular, they did not know how to write the square of equations---they simply added the '2' to the figures, not as superscript! Even worse, they called me 'professor'. Now that is objectionable from my point of view, since I have never had the good fortune to see the four walls of any university---never even had a full primary school education due to financial constraints in the jungles of colonial Africa where I was born and raised from 1938, working my way through life from the age of ten as a labourer; for this was an age when colonialism was in full swing and the ordinary man and woman in the colonies of black Africa, in the words of A.J.P Taylor, 'had no more history as if they were cattle'. Here is my article in full:-

"

FRIDAY, JANUARY 31, 2003 THE GHANAIAN TIMES PAGE 5

ADVERTISERS'S ANNOUNCEMENT

THE COMING REVOLUTION IN PHYSICS

1. We are all now invited to make suggestions for changes in physical theory, as it appears that physics is once again in the doldrums. Just about a hundred years ago, everybody agreed that almost all the difficulties in classical physics, some of which were spelt out by Henry Poincare at the 1900 Paris Congress of mathematicians, were swept away by Einstein's theory of relativity. In fact, Lord Bertrand Russell admitted in MY PHILOSOPHICAL DEVELOPMENT (Ch. 14), that everything he proposed in his Fellowship dissertation, later published as a book, was 'somewhat foolish' due to Einstein's theory of relativity (Ch. 1); yet he won his Fellowship on the merits of the ideas in his dissertation—on the foundations of geometry. There is a lesson in this story for science all over, in every science, as Russell himself pointed out in the same book: "Science is at no time quite right..."
2. Today physics is facing a quandary as grave as those that distorted classical physics before Einstein. Mathemati-cians are presently frustrated by the fact that every attempt to link the quantum theory to general relativity meets with seemingly impenetrable "Mathematical Roadblocks". As usual, problems in physics are not solved direct, except by what Einstein called

"a theory of ne principle", or "a theory of reconstruction".

3. What we now require in physical theory may be just a theory of reconstruction, as I cannot see how relativity or quantum theory can be completely falsified: there are too many proofs, including the bending of light in a gravita-tional field, the propagation of light without the eather, the photoelectric effect, and the ultraviolet catastrophic burst that never occurred or never occurs, to grapple with.
4. In my opinion, it seems that, as classical physics got into difficulties over the propagation of light, today's physicists are frustrated by the nature of time. Yet time is crucial because the quantum is time-dependent. If our time is our own in the absence of a universal time, can the quantum be universal—that is, since it is based on our own peculiar time system on this planet, this inertial frame? The technical term for the quantum is "energy second", often abbreviated to 'erg-sec'. How can we apply this to matter outside the earth? That is the first query.
5. Secondly, it is suggested that, time, as space-time, is purely mathematical, as stated by Russell: "SPACE-TIME, (sic) as it appears in mathematical physics, is obviously and artefact, i.e. a structure in which materials found in the world are compounded in such a manner as to be convenient for the mathematician".---[This is the opening sentence of Chapter XXXVI of Bertrand Russell's Analysis of Matter, but as I hope the reader will appreciate, intricate references and academic footnotes were or are

(generally) not appropriate in a newspaper article.]

6. Now then, again, the quantum is time-dependent, but how do we know that this new concept of time, space-time, which is purely mathematical, is the true nature of time to apply to matter in the entire universe, since mathematics is human in origin? Let me stress this point: "there is no longer a universal time" (Russell's phrase); and, also, according to Einstein, "There are as many times as there are inertial frames".
7. Are we sure we know what space-time really means? Being is time without mathematics. Time is being subjected to mathematics, for otherwise we could not have time in units; and without being, or consciousness, we could not speak of having time at all. For instance, in sleep or coma time is passing, but how much time? If we all went into coma, there will be no time. On coming to we would need mathematics to tell us how much time has passed, is passing, and likely to pass, using the theory of probability. Otherwise there is no time, only bland existence, as for the inanimate matter. Time requires points for the individualities of separate time units, seconds, minutes, hours, etc. Thus sentience, the ability to count, and arithmetic are required---in addition to the intellectual use of points. I have to explain that many of the terms used here are my own, for which I have written dozens of technical papers in explanation. If the reader does not understand, I plead that he or she may take them on trust.

8. The space-time equation is written as s2 =c2 t2—x2---y2---z2 [!!] based on the Cartesian co-ordinates, precisely as Minkowski stated. One important point is that the Minkowski space for time is flat, not curved. So why are scientists always calling time "curved space-time"? I see this as part of the wrong interpretation of space-time.
9. Minkowski stated clearly that space by itself and time by itself cannot exist, and therefore we must call both 'space-time', using the Cartesian co-ordinates. It means time is space divided by points to get time-like intervals in any space. When these time-like intervals are mechanized in the clock, and then applied for all activities and events, all space and matter, then, in my view, the time becomes independent of the space it is being applied to; but the space cannot be independent of time because it takes time to traverse any space. The notion that time is not independent of space under relativity theory must be discarded. It is independent of space, but not in the Newtonian sense. Time cannot be gained except by the application of points to space, therefore time is space-time, or space-timed; but once that is done, the time that is gained, and which is then applied to space, is independent of space---the clock is independent of any space, yet it is applied to any space.
10. These topics have been discussed in two small monographs that are soon to be published by AFRAM PUBLICATIONS in Accra (I hope it will be very soon). One is called THE MATHEMATICAL THEORY OF TIME, as sketched above. The other is

asking the question HOW OLD IS THE UNIVERSE IN TIME? They are philosophical in tone, but obviously have implications for the coming revolution in physics, which I believe will revolve around the nature of time, mainly because of the quantum; and so if we want to link the quantum to general relativity, then we need to take a good look at the nature of time.

11. This advertisement has been placed to give prospective readers an idea of the contents of the two little books, some of whose conclusions are that:
    a. All time consists of units, derived from space by means of the Cartesian co-ordinates, or points, for otherwise we could not have the individual time units we know---like the second, minute, hour and so forth, all the way to the year, which is also one unit of time.
    b. All time units on this planet are derived from the earth-year as fractions thereof. There is no other sense of time on this planet. The year is itself one unit of time. It has to be repeated to pass by and make time continuous---thus the continuity of time is achieved by means of the succession of time's units. The year increases in numbers to pass by, together with its fractions, or every other unit of time. The clock ticks second by second, in the absence of that there is no continuity of time. If a year ends and another year does not begin imme-diately afterwards, where will we get the next second from?

c. The supposition that the earth-year is basic to our time system automatically makes the year our basic SI of time, not the second, since all other time units are derived as fractions from the year.
d. The above proposition is best illustrated by what I prefer to call. "The time Paradox of the Lone Survivor", which goes like this: suppose that the earth is destroyed through collision with a massive asteroid, but one person survives, floating on a comet-like lump of rock. He had no time to begin with, but later finds an old mechanical clock. It has stopped, and he winds it up again, and it begins to tick away in seconds, as it is meant to do. But how is he going to set it for minutes and hours, days and weeks, months and years? The earth is gone, so how is he going to get his time units derived originally from the earth-year? We should therefore regard the earth-year as our basic SI of time out of which all other units of time are derived.
e. As the year is our basic SI of time out of which all other time units are derived and mechanised in the clock, there can be no time dilation unless the year dilates. We are often told that time dilates because clocks perform differently (faster or slower) under certain conditions. Yet the clock is not the whole of time; it is based on something else—i.e. the year. So a dilated clock is seen to have dilated only by comparing it with 'a standard or correct

clock', being one that is running correctly with the year and its fractions. Thus a mere dilated clock is not "time dilation" as a general thing applicable to earth time, unless the year itself dilates, by the earth running faster or slower.

f. Furthermore, warped space is tenable only under Riemannian geometry, and as such, has nothing to do with the Minkowski space which gives us time intervals by application of points to space. In fact, the Minkowski idea is better understood as t=s/p. Time is space divided by points. Therefore the often repeated concept known as "curved space-time" is completely wrong. The Minkowski space for time is flat. The clock's face is flat.

g. As mentioned above, the continuity and passage of time can be logically explained as the succession of time units---for after all the year itself increases in numbers to pass by and make time continuous."[5]

---

[5] **COMMENTS:** I will not reject any of these ideas as entirely wrong, but rather would phrase some things differently. For instance, my theory that we do not know anything about time except how it passes has now been fully developed. In this article it was merely implied with the assertion that "the continuity of time is achieved by means of the succession of time's units". Today, I write that the units of time constitute the time metaphysically, so that the passage of the units, say, of the years, is the passage of time, because that is all we can ever know of time, namely how it passes. Also, the statement "Being is time without mathematics" will stand, since mathematics plays a crucial role in the having (or creation) of 'mechanisable' time for the clock. Without mathematics we

---

could not have created the clock. But the actual definition of time as, "Time is being subjected to mathematics..." would be expanded to "time is being subjected to mathematical (or repetitive) cycles"---the year for instance, because it is mathematically produced with the meridian point and the space traversed round the sun, so that a second is a specific unit of space, and the full year is only a cycle, repeated to become years continuously. Hence the arrow of time idea is no longer needed. Finally, a time system consisting of units is discrete time. Discrete time cannot march. History is not the march of time but of events with their associated dates of occurrence. Discrete time cannot curve either, so that the concept of 'curved space time' is logically untenable, and so forth.

CHAPTER 2:

Discrete time, and why it matters a lot

The actual thesis of this little book is discussed in the next chapter. Here it is only necessary (briefly) to outline the sort of time that is philosophically conceivable on the Einstein notion of time. The chapter is shorter than the one following because, in essence, the Einstein idea of time is simple, as he himself put it.[6] It can even be stated in a few sentences, and no mathematics is required. But we have to go into a little detail so as to justify why it is different from the Minkowski time of four-dimensional space; as well as showing why the Minkowski time is logically unacceptable, or untrue for short.

In the very strong words of Professor Sir Arthur Eddington, who was one of the handful of world authorities on relativity, after Einstein's researches about time we haven't got "true even-flowing time".[7] His opponents are described as making "meaningless noises".

---

[6] The bulk of the chapter is devoted to comments on some of the many myths and legends of time, but the actual outline of the theory is very brief indeed.

[7] His actual words are quoted below. It is the first admission that the Einstein notion of time is different from traditional time. I identify it as 'Discrete Time' that has very serious implications for technical philosophy, social theory (otherwise known as Ethics) and religion, since it can be used to undermine most of the teachings of all the religion about the nature of time and life. For it must be understood very clearly that how we live our lives or must live our lives is mainly dictated by the religions, so it is imperative that they do get their ideas about life and time right. The life is 'given', we just have it. But the nature of the time, which is debatable, determines the cultural life of man, usually in an oppressive fashion without escape. Culture is the eternal curse of man from which he has no escape. If the milieu is good and spared of natural calamities, the life will be

In effect, the Einstein notion of time makes earth time necessarily discrete, being the time we ourselves help to create with points in application to space: sentience is require for counting the orbits of the sun as 'years'. Bertrand Russell has also confirmed my opinion that the Einstein adoption of the Lorentz local time as "time, pure and simply", means he knew that space-time was implied in his idea of time, long before Hermann Minkowski made it explicit with his own brand of intricate mathematics, which vitiates his theory because the basis of it is imaginary time co-ordinates----and you can't do that in logic. Yet all mathematical deductions are (and to be in accordance with reality, *must be*) based on a sound logical foundation.

The whole new idea of time began to take shape when Lorentz discovered his time dilation, otherwise known as t^1, or local time, a time system that was different from the one in normal use; the first occasion that anybody had managed to create a different sort of time that was neither general nor absolute. I have no doubt that calling it 'local time' disturbed Einstein a great deal. For he had conceived reality as consisting of four dimensions: three of space and one of time. His own formula was the 3+1 idea. It meant that man had to add the time to phenomena, but whose time, what time or which time, if time has been discovered by Lorentz to be changeable and not absolute? The result was his theory that time is essentially dynamic and not absolute. After thinking about the matter, perhaps a great deal, he came to this conclusion, as he put it: "All that was needed was the insight that an auxiliary quantity

---

good; if it is hostile and the Gods too are angry the life is unbearable. Traditionally, religious and political leaders have always combined to rule man's life woefully, particularly in human sacrifice, slavery and wars.

introduced by H.A Lorentz and denoted by him as 'local time' can be defined as 'time, pure and simple'."[8]

Thus, logically, every time becomes somebody's local time. It means there is no standard frame (or reference) for time: a second here cannot be the same as a second anywhere else. A revolutionary idea of incalculable consequences. We are still trying to understand it properly.

The best way to understand this new concept of time, I think, is to compare it to some of the myths and legends of normal or traditional time, viz:-

1. ETERNAL TIME

We were told that time is eternal, just because it is always there: it was there when we were born, and there as we die. Yet the Einstein new concept of time Professor Eddington was defending is not eternal but limited to an inertial frame. According to Abraham Pais, under relativity, "There are as many times as there are inertial frames. That is the gist of the June paper's kinematic sections, which rank among the highest achievements of science, in content as well as in style. If only for enjoyment, these sections ought to be read by all scientists, whether or not they are familiar with relativity. It also seems to me that this kinematics, including the addition of velocity theorem, could and should be taught in high schools as the simplest example of the ways in which modern physics goes beyond everyday intuition..."[9] Strong words. The suggestion puts

---

[8] Quoted by Abraham Pais, in his Biography, *Subtle is the Lord...*" Ch. 7. The author also gives the background discussions between Einstein and his friends before he discovered his theory of time in the same chapter.

[9] Abraham Pais, ibid, Ch. 7.

scientists and philosophers ignorant of relativity to shame.

My reply to the concept of eternal time is that discrete time cannot be eternal. According to Professor A. N. Whitehead (Philosopher, Logician and Mathematician of genius): "A time system is a sequence of non-interacting moments [However the moments are defined]."[10] For my part, I try to avoid such vagueness by defining time as a procession of time units. This is because the basic unit out of which all other units are derived is the earth-year, but it is only one unit of time that has to be repeated for time to continue. So the years are in procession, and their sub-units, or fractions (from the second upwards) have got to be in procession too. The reader can see that in this simple definition the problems of the passage of time and its continuity have been solved without any need to rely on the mythical arrow or arrows of time. In other words, perpetual time arises from the procession of time units. We get the units with the application of points (or mathematics) to the basic unit, the year, which is also one unit of time based on the meridian point or line. These units of time are always in procession: second, second, second; or year after year after year, hence perpetual time. This suggestion can be seen as true when we apply what I describe as "The time paradox of the lone survivor", as mentioned in Chapter One above, namely without the earth and its orbits, you can never set any clock to give you time in consonance with the features of the earth, which means you could not live on this planet in a logical, sensible or wise manner.

---

10 Professor A.N. Whitehead in *The Principle of Relativity*, Ch. IV, Cambridge, 1922.

DURATION

1 (a) Something must be said about the sense of duration under this heading. Duration is the internal time-sense that tells one that a second is shorter than a minute, and so forth. But the duration is derived from the basic unit of time, which is the earth cycle; for it is the earth-year that gives meaning to the units of time we use. Once the year is known, and sub-divided, each fraction comes to exist in the mind---i.e. as shorter as or longer than this or that---a distinct and independent unit of time in its own right.[11] The human brain is highly malleable that is why we can acquire skills simply by training the brain-mind system to do tasks. The mind absorbs impressions as sponge absorbs water, and they stick. It is said by anthropolo-gists that almost all impressions from the external world since the dawn of existence (or of life) are preserved in the mind, hence the unconscious mind's grip on us. Thus once we create the basic unit of time by using external cycles, the mind just uses the units based on them as units of time, as required, as divided or otherwise. According to Professor Eddington, "The rough measures of duration made by the internal time-sense are of little use for scientific purposes, and physics is accustomed to base time-reckoning on more precise external mechanisms."[12] Because of the adoption of the Einstein notion of time in philosophy (or the philosophy of science), this scientific view of time, as a union between external cycles and the sense of duration, is identical to those espoused in philosophy, rational philosophy or the philosophy of

[11] This has been a source of confusion and mysticism in the past as people failed to notice that every unit of time is a fraction of the year and not a unique creation of whoever gave us time.

[12] *The mathematical Theory of Relativity*, Ch. 1.8.

science.[13] They are units of time, and I have discussed how they are secularly created requiring points, the intellectual use of points and therefore sentience; they are of various lengths; and the mind is trained to notice the differences. So that if you ask somebody to wait for ten minutes, the person will know automatically (mentally) that it is shorter than asking him or her to wait for ten hours.

2. TIME ZERO AND THE DAWN OF TIME

In their efforts to explain the existence of time, even after relativity (which implies that they reject the Einstein notion of time, a hanging offence in my books!), some scientists have invented what they call "Time Zero". All that needs to be said is that the above explanation of time makes this new theory redundant. And I want the reader to know that putting my opinions in this kind of lofty language is being extremely charitable. Bertrand Russell would have put it much stronger than that. He used to describe some ideas as "the most fundamental of my beliefs is that so-and-so is rubbish". The popular phrase in astronomy, known as 'The dawn of time', is also rubbish. To make any sense it should be 'The dawn of existence'. Time originated from this earth---after it was formed and began to orbit the sun; even then time had to wait for man to appear on the scene, because time requires points and the ability to count the orbits of the sun as 'years'---or there will be no years and no seconds (and the rest it) that are obtained (only) from the breakdown of the year. The practice of equating any motion with time is not logically defensible because, to know 'how much time' (through the creation of 'quantified time' as discussed in Appendix I below), you need a clock; and to invent or create a clock you need a union between the sense of duration and repetitive

[13] One has to be careful because the religions, crooked Asian gurus and mystics also espouse 'philosophy'.

external cycles—otherwise the mere physical orbit of the sun will not become 'a year'.

Thus time cannot be just any act of motion or 'Being' in itself *per se*. It is true that in sentient Beings, just being there means 'continual ageing', which shows time going. All the same, to know 'how much time' (or to demonstrate that the time is going at all), you need a clock which cannot be said to be inherent in the process of ageing. The mistake in that discredited line of thought was the implication that time is 'marching' throughout the entire cosmos---and the same everywhere, not as something that varies from frame to frame, so that how it is created in each frame is the essential problem of time. And let me explain that the best logical solution of how it is created from frame to frame is the supposition that we use external cycles repeatedly and call them time units---the years, for instance—so that sentience is required.

2 (a) TELLING THE AGE OF THE UNIVERSE

What temporal yardstick can we us for telling the age of the universe accurately? This question will no doubt be regarded in scientific circles as a heresy originating from someone infernally ignorant of counterintuitive mathematics. Be that as it may, one cannot prevent ridicule from vain intellectual snobs from 'The Great Universities', but telling the age of the entire universe by earth time is not so great, though. Actually it is illogical and unworthy of those who propose it. First of all, I don't think it is really necessary to worry so much about telling the age of the universe. It is so vast; the nearest star likely to have earth-like planets is so far away it will take millions of years to reach it. Even then, what are we going to do there---sell some of our bad habits to ruin their lives as we have managed to ruin life on earth? Secondly, the age question---what for, and how do you estimate it accurately? We think we are important, yet listen to this, as Professor McNeil Dickson put it in is book, The Human Situation: "There are billions of stars so immense that millions of our petty sun will find room in one of them".

Instead of spending so much money and intellectual energy speculating about the age of the universe for no conceivable (or reasonable) purpose, use, or benefit to mankind, I think we would be well advised using our resources for guarding the earth from collisions with massive asteroids.

Normally the age of the cosmos is told in earth-years---what a joke. How long in temporal terms is the year? Nobody can estimate that simple thing with which to approach the estimate of the age of the cosmos with any accuracy. Furthermore, has it never occurred to these wasteful, insensitive and snobbish academics that the sun came first; and it was a long time after that before the earth came to be in existence, (no years to cite because the earth was not there orbiting the sun!) They themselves tell us that it took ages after all this before the earth began to orbit the sun regularly, have its moon which influences conditions on earth, and generate any kind of atmosphere in which man could thrive. Then God made the mistake of creating man to inhabit the earth and bring with him drug addicts, bank robbers, corrupt politicians, mad scientists, teenage pregnancies, criminal immigrants, wars, wars and more wars, burglars, lawyers skilled in the dark arts, arrogant judges, religious gurus, paedophile clerics, Presidents, Queens and Kings, and so forth, all of which make life on earth something of a curse, even without taking account of natural disasters. Thirdly, our sun is so small that going round it does not take much time; therefore the earth-year is not, and can never be, an adequate temporal yardstick for estimating the age of a universe so vast that we cannot even imagine where it ends. All that nonsense of the age of the cosmos being about 15 billion years is just so much hot air from overfed scientists.

3. TIME AS PART OF CREATION

We are told that time was created by God when he created life on earth. This, of course, is religious nonsense, and needs not detain us much. No evidence is produced, and therefore it is not a sensible topic for discussion. Besides, great men like Charles Darwin have wasted their valuable time to refute the concept of creation logically and convincingly. Any person who still believes that the world was created by God at Noon in AD 4004, as Archbishop James Ussher told his credulous audience, needs his head examined.

4. ABSOLUTE TIME AS OPPOSED TO DISCRETE TIME[14]

Of course, this whole book is discussing the sort of time that is conceivable to imagine under the theory of relativity; and it is argued that such a time system must be discrete. We look around and find that we actually have only discrete time in operation on earth---and that it had been so long before Einstein was born, therefore his theory after all was pragmatic. I will not insult his blessed memory by claiming that it was not original. It was highly original because so many myths and legend have made time seem more mysterious than the creation of the universe itself.

---

[14] The question is whether time is absolute or discrete---the exact opposite of absolute time. This is interesting. For discrete time is bound to have quite different implications from that of absolute time, as we have seen in this book. This is so serious that I hope people are surprised by the consequences because, of course, nobody knew it existed to have any idea of what it may imply. Being so serious a topic, I am not surprised that initially people reacted to my ideas with more hostility than they showed to Einstein---at least he had PhD!

So, then, we believe that time under relativity is necessarily discrete because it is created on this planet with the application of points to space otherwise we could not have it in units---and we do have it only in units, since time is known and used in units alone.

Discrete time cannot permeate the cosmos 'generally' as an *absolute entity* so that a second here is the same everywhere else. This was specifically proved by the researches that Einstein carried out more than a hundred years ago. It is a subject that is, therefore, no longer open to question, as Professor Sir Arthur Eddington has confirmed.

The difficulty arise from man's acceptance of the consequences of discrete time, as many of the conclusions that we can infer from the having of discrete time on earth are anathema to the human mind and religious beliefs. Technically, discrete time does not exist in nature at all; it is the product of points, so that a theory of numbers is necessary, the ability to count is required, and sentience is obligatory: somebody must be there to count the orbits of the sun as years or there will be no years and no seconds. All this makes time absolutely secular, originating from this earth, and limited to it---whether you and I like it or not, is the question. Personally, I do; but I don't know about you. However, those of us who accept it, point to the benefits of having a rational and logical time in consonance with the essential features of the world. Those who reject it are hoping that there is God and that he has prepared luxurious life for all believers after death---so that they can die happily. True or not, nobody can find out. The dead never come back to report on their experiences after death. Those who accept the Einstein theory of time claim that they are more intelligent than those who reject a theory with such exemplary scientific credentials; yet that, in the end, is a matter of opinion since it is judgemental.

5. CURVED SPACE AND CURVED SPACE-TIME

In Einstein's theory of gravitation space is curved, and that is what causes all gravitational pulls. So gravitational pull is not conceived as 'an attractive force' such as we know between atoms or the elements. It is rather conceived as some form of 'a slipping slope' due to the curvature of space. And I am happy to announce to the non-professional that the idea is no longer a theory: it has been proved. It was our own Professor Sir Arthur Eddington who led the group that confirmed the theory---after which confirmation it became a fact of nature.

You may ask why time is conceptually included in the curvature of space when it obviously is not so in physical reality. The reason is to be found in the contribution of Hermann Minkowski. He once gave a lecture about relativity known as Raum *und Zeit*. In German (it means Space and Time, but if you fail to mention the German version the academics will look down on you as ignorant!) His lecture was particularly welcomed by Einstein and his handful of followers, because at the time, the world's scientific establishments were ignoring relativity and Einstein. Unfortunately, as time goes on, it appears that Minkowski rather succeeded in the permanent distortion of relativity with his equation of space to time.

I quote how he began his lecture below (again, in the absence of that some scientists will sneer). I have read it a thousands times; it seems different from one writer to another. Then I saw what is quoted by Professor Eddington in his book Space, Time and Gravitation (Chapter 11), and accepted that his version seems the more accurate reproduction of what Minkowski actually said. This is it: "The views of time and space, which I have to set forth, have their foundation in experimental physics. Therein is their strength. Their tendency is revolutionary. From henceforth space in itself and time in itself sink to mere shadows, and only a kind of union of the two preserves as independent existence." (Hermann Minkowski, Raum und Zeit, Cologne, 1908.)

After this Minkowski lecture and subsequent technical papers, the whole of science were converted to the view that space and time constitute one entity; so that when space curves it takes time with it. One of the bizarre inferences from this idea is that time travel has been scientifically 'proved'.

Minkowski dominates all theoretical physics, cosmology and astronomy. Indeed it can be argued that he looms large all over the pages of this little book as well---see Appendices III & IV below. For now, it is only necessary to point out that the Minkowski theory which equates space to time is not regarded in philosophy and logic as successful because he based his formula on imaginary time co-ordinates; and that, even in mathematical physics, they ought now to take notice of the judgement of Professor Eddington, to the effect that his formula is fictitious---and also of Bertrand Russell himself, a philosopher a founder of the philosophy of science, who said the formula is compounded for the 'convenience of the mathematician'. Even school boys will argue that this is not the best way to determine the nature of physical reality. Unfortu-nately, scientists adore the Minkowski theory because, secretly they detest the 3+1 formula as not entirely objective, since man is the one to add the time; and yet Einstein has shown that there is no universal, or god-given cosmic time, thus making man supreme in the determination of physical reality. So the Minkowski proposal that enables them to pretend (it is only a pretence because it is not true, even Eddington told us so), is preferred. '*S=ct...*', as Russell has averred, makes things easy for mathematicians, but it is not true of the physical world (the same Russell said it is an artefact); therefore the concept of curved space-time and time travel based on it are all bogus science. They will never give me any honours for pointing this out;

but, at 71 going to 72, I do not need honours from scientists who peddle untruth as a scientific discovery, not even from the Royal Society.

6. THE TWIN PARADOX, TIME DILATION AND THE CLOCK PARADOX

These three topics are lumped together because they are what underpin the current distortion of physics. The twin paradox is the silliest. It assumes that time travel is a reality, and therefore if one twin travels by time he would return to find his other twin aged considerably than himself because during time travel physiological processes are slowed to make the time traveller age slowly. This is how one writer puts the idea: "...suppose you leave the earth in a very fast space ship, journey through space for seven years, and then head back to earth; the equations of special relativity predict that on returning home, you will find that everyone will have aged not fourteen years, as you will have done, but fifty. According to your earth-bound friends, time really will have slowed down for you..."[15] There are no such equations in special relativity—at all. What Einstein said is quoted below, and it concerned only the clock paradox which he stressed had no bearing at all on relativity, and cannot be used to contradict it. What these scientific mysticisms are all about are these, (a) Time dilation and the clock paradox are assumed to slow time at high speeds; (b) the Minkowski formula is assumed to predict time travel; since space is curved, and curved space-time is posited, shallow thinkers

---

[15] Robyn Arianrhod, in her book, "EINSTEIN'S HEROES..." Icon Books, 2004, p.182. Books about time travel are now all over the place. Publishers are intrigued by these bogus scientific assurances that time travel is feasible---"as predicted by Einstein's theory of special relativity". It is a false claim. They are basing their ideas on Minkowski not Einstein, but they know that publishers go potty when Einstein's name is mentioned.

conclude that when people travel fast their physiology will slow down because they are in space, and time dilation implies that since they have to travel fast, their physiology will slow down---all of which amount to sheer mystifying, contradictory nonsense, so annoying that one is left to feel sorry for mankind, in the knowledge that man will always succeed in falsifying rational theories and revert to superstition, religion, mysticism and cant.

Let me stress this with all the strength in my veins: there is absolutely no shred of evidence that time travel is feasible other that the discredited Minkowski equation of space to time. Also it is not proposed in any theory (scientific, mystical, religious or philosophical) that moving fast even at the speed of light could slow the process of human growth. There is no connection between speed and physiological growth or decay, none whatsoever. They normally base such theories on the concept of time dilation or the clock paradox, both of which are discussed immediately below---actually they do not affect human physiology.

Next, time dilation. This is a true scientific theory, discovered by Lorentz, but it only says that a moving clock would be seen by those outside its medium of travel as running slowly. Scientifically it is known as the dilation of time as a measure of moving clocks. The name is wrong. It is wrong because all time will not dilate, only the one travelling clock will be seen by outsiders as running slowly.[16] Those carrying the travelling clock will notice absolutely no difference in its performance. So Lorentz called it "Local time" or t^1 and put it aside. As far as he was concerned it was not the true time, as Abraham Pais put it: "He proposed to call t the *general time* and t^1 the *local time*. Although he

---

[16] This is another poignant example showing that science will always require philosophical interpretations.

did not say so explicitly, it is evident that to him there was, so to speak, only one true time: *t*."[17] As mentioned in this book, Lorentz claims he failed to discover special relativity because he did not attach due importance to his own discovery that time can begin from anywhere---for that is the phenomenal importance of his discovery, not that time (normally) runs slowly with speed. The very idea of having a different sort of time from the one normally in use destroyed the concept of absolute time, which was worrying. For time in no way runs slowly with speed or without speed, unless it is conceived as intertwined with space, as Minkowski supposed, and which notion we now regard as false.

Thirdly, we hear of the clock paradox as the real and irrefutable scientific reason why (as discovered by Einstein himself), speed slows physiology. In fact, this is what he said, according to Abraham Pais in his Biography already cited: "Einstein rather casually [meaning not very seriously] mentioned that if two synchronised clocks C_1 and C_2 are at the same initial position and if C_2 leaves A and moves along a closed orbit, then upon return to A, C_2 will run slow relative to C_1, as often observed since in the laboratory. He called this the result of a theorem [meaning a curious fact yet to be explained, but certainly not a scientific 'thing'] and cannot be held responsible for the misnomer clock paradox, which is of later vintage. Indeed, as Einstein himself noted later [E16, the original source, always given by the author] *'no contradiction in the foundation of the theory [of relativity] can be constructed from this' since C_2 but not C_1 has experienced acceleration*'." (My italics.)

[17] Abraham Pais, ibid, Ch. 7.

6 (a) GRAVITY AND TIME

Yet still there is another mystery about time, namely the effects of gravity on time. Note that all these mysteries are supposed to return time from the secular (as can be inferred on the Einstein notion of time) to the mystical world of religion and God. We are told that gravity slows time, and that this has been proved. In fact it cannot have been proved because gravity, conceived as a mere curvature of space, cannot have mysterious tentacles to affect all time---not even one clock, but all time *per se*. Here are the true scientific facts (and I quote from Professor Jeremy Bernstein's book "ALBERT EINSTEIN and The Frontiers of Physics", Oxford, 1996, pp 110-111.): "In the absence of gravity, space and time are distinct entities. In the metric of special relativity they play distinctive roles. But in the presence of gravity the metric is altered, and space and time become mixed up with one another. The metric has four coordinates, but the space and time coordinates become entangled. Only when gravity is weak can they be distinguished in a useful way..." Gravity distorts the metric of space (not space-time, mind you, because we have explained that the Minkowski formula for it is "arbitrary and fictitious"), as Professor Eddington put it. Once we dismiss the Minkowski 4-D geometry, gravity cannot affect time since time is not linked to space, especially in a metric such as the earth's where gravity is not as strong as we suppose may be happening in a black hole.

In special relativity space and time were separate. There are two reasons for this. One is that the metric on earth is not a 'strong gravitational field' where time is *supposedly* distorted by space. So the question does not arise at all. The second is that, obviously if the Minkowski formula for equating space to time is fictitious, then space and time remain separate from one another, so that however strong, gravity cannot affect time. Scientists have invented another theory about time's slowness in a gravitational field (about which some theorists are madly excited). It is called 'Singularity'---a situation where gravity is so strong that space and time are fused, or that time is effectively obliterated. Yet it all depends on how we define time; for if time is not defined as in the Minkowski four-dimensional space, or 4-D geometry, where space and time constitute one entity, then time cannot in any way and manner, shape or form, be fused with space, because it remains independent of space. In any case, there is nobody in the black hole to invent his time units, i.e. by counting the orbits of some other body as 'years', and so forth. Time is not 'Being' on its own. Just 'Being There' does not give you time; trees have no time that is mechanised in a clock. Time is also not just motion as some scientists tend to regard it. Time requires points and the ability to count the intervals between points as 'time units', or 'time moments', as Professor Whitehead called them---the years, for instance. I must also point out that Russell too defined time or space-time as 'relation between points'. Altogether, since Einstein, logicians and philosophers of science have formed notions of time that seem to be identical, rather surprisingly. Formerly we used to think that no two philosophers could ever agree on anything. Furthermore, the anti-intellectual (thoroughly childish) sneer that no philosopher has ever solved any problem conclusively has been proved wrong. Philosophers have shown how the problems of the passage and continuity of time are easily resolved conclusively without the mythical arrows of time by way of the Einstein theory of time which gives us discrete time. It was the same problem that the religions called 'perpetual time'. A concept in which they used to take great delight because they thought it is so mysterious that nobody could ever resolve it with logic or mathematics without recourse to divinity. But in fact, if time is discrete, as secular time originating from this planet is bound to be, then, as we see the year increase in numbers to pass by, the units of time in procession is perpetual time, 'pure and simple', as Einstein would put it.

7. THE QUANTUM AND TIME

The technical name for the quantum is 'energy-second', as mentioned above. If the Einstein secular notion of time is correct, then the energy content of the quantum is natural, but the time period for its occurrence is peculiarly our own. This would seem to suggest that the quantum cannot exist in the cosmos at large in exactly the form we know it. This is yet another serious implication for the notion of discrete time. We can of course work out with mathematics the necessary time unit in another time system equivalent to the earth unit for the quantum. But that is another matter. As things stand, we must assume that the quantum is peculiarly our own 'ultimate' unit of energy, not known to be capable of existing in exactly the same form elsewhere in the universe.

7 (a) BEFORE AND AFTER TIME

The Meridian Line and The International Dateline are both meant to refer to the line or point where the present year ends and the next year begins. 'Before and after that line' there is no time, in so far as we rely on the earth-year, or use the earth's orbit of the sun for time. _No orbit, no time_. We have no time system outside the earth's orbits. The subunits of the year we have obtained from past experience are often used as if they have independent existence in nature outside the year. It is a mistake that has grown out of repeated experiences of using and subdividing the earth-year down to the seconds. To understand this, let us assume that we are now using the earth-year for the first time, without knowledge of how it can be subdivided down to the other time units, from the seconds upwards. In that scenario only the year will be known. But having used the year for so many centuries, we have learnt to subdivide it; therefore we count the seconds and the other units of time to make up the year---before and after the year. But logically they do not exist. There is no time before the year and no time after the year.

Often we are told that we can see or 'experience' existence before time and also after time. We are told that, properly positioned, we can see the future and the past: that is, before the Meridian line, and after the Meridian line. 'Before' the year by way of the Meridian line is said to be looking ahead into time or what is waiting to be reached with the march of time; and 'after' the line is looking back into time or what has come to pass, going back several centuries; never economical with numbers in this area, only 'economical with the truth'! They claim it can go as far back as to the time your grandparents were born, and so forth. It does not make sense to me because the line is artificial; it does not exist in nature physically; and, in any case, discrete time does not and cannot march.

Rather, we can speak of 'before' and 'after' events, not 'before' and 'after' time. The coming or the past year as seen from suitable positions is, logically, the 'coming' or 'past' event or events. They will take time to come round, but that has nothing to do with time because 'being' is not time. The time it will take for the event or events to occur cannot be known by simply looking at the events; it can only be known through what I describe as 'Quantified Time', as explained in Appendix 1 below. Also, the eye-view showing both states counts for nothing. In logic it is just a view divided by yourself into 'before' time and 'after' time in your own fashion, scarcely a matter of importance. Time is so spoiled with legends that we have always to be on our guard and scrutinise concepts of time carefully. For instance, since events occur on the chemical, atomic or quantum levels of reality, we have to reckon what happens as millions or even billions of events, and simplistic to just call them 'before' and 'after' time, without even being able to define what is time. The normal intelligent man's definition equates 'Being' with time; yet 'Being' on its own is not time. You need points to create 'quantified' and therefore usable and quotable time. That is the only way to tell "How much time"; and my argument is that all we can ever know of time is the ability to tell "How much time". The true nature of time can never be known, but we can have a guess, and the field is open to suggestions, on condition that they are strictly logical or philosophically acceptable.

Yet still some mathematicians have concocted intricate theories about time whereby it is claimed that we can see into the future and also far back into the past by using the Meridian or International Date line. And Let me explain that, once the earth-year is relied upon as our basic unit of time, the International Date Line or the Meridian Line is implied. I don't think many theorists realise that, using the earth-year (and its subdivisions, from the seconds up) in all their suppositions, as they do (there is nothing else to use) means when they talk of looking into what is there before and after time, they are referring to seeing what is there far ahead of the year, or far back into the past years, since the year is the beginning and end of our earth time. But all that is fantasy---together with associated liturgy, rituals, incantations, religion, taboos, mathematics and so forth---because the year does not even exist in nature. Man manufactures the year out of the pure physical motions of the earth. Without sentience to count the orbits of the sun as 'years' there will be no years and no seconds and all the rest of it; there will be no quantified time only 'being'; but 'being' on its own does not constitute time, because time requires points. 'Being there' is purely a matter of physical, organic and physiological chemistry, inertia and motion.

Another important point is that accepting the Dateline means we are using only 'discrete time'. That is to say, our time is necessarily discrete---consisting of units of separate and individual units, or moments, as Professor Whitehead put it. We have to have successive years for time to continue. So the year is the basic unit. Before the year there is no time, and after the year there is no time: discrete time is ended when the unit is spent. Our time is ended when the year ends. We have to have another year for time to continue, as I have said. It is necessary to repeat this because it is crucial but often overlooked by theorists.

If the earth does not go on to orbit the sun again, our time will automatically come to an end. Discrete time does not exist outside its unit. You cannot see before what is not there; similarly, you cannot see after what has not 'yet' occurred. What we do to get our concepts of 'before and after' may be called 'anticipation' based on history. We have to rely on past experience for both concepts of 'before and after time' even before the time has occurred. After all, 'Existence' *per se* is not time. You cannot mechanise bland existence for reproduction in a clock; there simply is no logical or mathematical mechanism for it. But by using points to divide space we can create 'relation between points' (according to Bertrand Russell), as time units---the principal example of which is the earth-year which is created from the Meridian point—and such units are, of course, eminently susceptible to mechanical representation due to the involvement of points. This is the definition consistent with the Einstein notion of time as 'local time' derived from local space, or technically, 'space-time'. And, as I have pointed out elsewhere, it is actually pragmatic. We had it already in use before we discovered the logic of it, since that is the basis of earth-time as we get it from the earth's repetitive orbits of the sun. There can be no system of time without the same logical credentials. We can even speculate that every sentient beings in the universe will have a time system similar to our own---that is how important Einstein was in philosophy apart from his numerous scientific discoveries. Albert Einstein was simply unique and incomparable.

Now, although technically impossible since time is bound-up with existence or 'Being', but if you can have any experience before the meridian line, you are not seeing 'before time' because the time has not yet materialised. 'Yet Being' on its own without somebody to count the orbits of the sun as 'years', is not time. Time requires points for the creation of cyclical units of duration so that the time's duration is known in the mind---only that is time; the year, for example, as produced by the Meridian Point. The subdivisions of the year too are obtained with points. So until the year materialises we have no time, since all other units of time are derived from the breakdown of the year, thus they can only be defined in reference to the year. Otherwise time is indefinable---the year, on its own, can never be defined logically. All we can do is to create units of time by means of regular or repetitive motions or cycles; so until the cycle is created (or completed) the time cannot exist; but once it is reached, it is gone: you cannot experience existence 'before' what does not exist. But time does not exist for us; it is passing, since that is all we can ever know of time. And being a system of discrete time, once the unit is created, it is gone. The year is ended when we reach 31st December.

In seeking a logical explanation of time, the units or fractions of time obtained from the year with points (which means all that we know as 'time', from the seconds or cesium units to the hours weeks and months) are disregarded, or must be disregarded. The cesium units are included because they have to be based on the second to make temporal sense. I know that atomic time units are often treated as metaphysical entities unique in nature, but that is wrong, unless their reliance on the second can be avoided, which is impossible. So let us say you know that an event will occur ahead at 10.am, and another one at 11.am, one hour after that. These will normally be called 'before' 10.am, and 'after' 10.am. But logically it is wrong to say 'before' the time 10.am. It is time all right; but it is not natural time. It is a fraction of the year. What you imply is that the events will occur at so-and-so fractions of the earth-year. For these units of time do not come into existence on their own---they do not exist at all but as fractions of the year. Therefore they are not events 'before' and 'after' time as is normally supposed to cause confusion in the logical interpretation of time. All theories of time are replete with similar mistakes. There is only one time, the year. Even atomic time is based on the second, and therefore based on parts of the year. We can use some other repetitive cycles rather than the year for our time, but then the same problem will arise---there can only be one time overall applicable to this planet. Furthermore, it is not correct to treat time as a naturally objective entity existing out there on its own and capable of complete mathematical representation simply because that is not what time really is---not at all. It rather originates in the mind. About ninety per cent of it is contributed by the sentience that counts the orbits of the sun as 'years' and help to breakdown the year down to the seconds.

CHAPTER 3:

The application of time (as opposed to the nature of time)

THE PROPOSITION

I cannot query the notion that time can be equated to space, or that by the Minkowski formula it has actually been achieved. I do not know enough mathematics to be able to question scientists about that; but I like to point out that the Minkowski theory is regarded as (a) Arbitrary and fictitious at least by two great thinkers who know enough mathematics;[18] (b) It is compounded merely for the convenience of mathematicians; and (c) it is my believe that the research into Order and Simultaneity through which it is

---

[18] Of course, the Minkowski attempt to show that time is the same as space in a world of 4-D geometry was not successful, as Russell made plain in this quotation: "...the philosopher cannot but feel dissatisfaction with the apparently arbitrary assumption about intervals...there is great difficulty in suggesting any non-technical meaning for interval; yet such a meaning ought to exist, if interval is as fundamental as it appears to be in the theory of relativity." (The Analysis of Matter, Ch. XXXVIII.) It may seem as if Russell's strictures were directed against relativity as a whole. On the contrary it seems so because Einstein had adopted the Minkowski proposal equating time to space and space to time; but in reality it was Minkowski who introduced the concept of intervals. Let me add that even though it is true that eventually Einstein was persuaded to incorporate the Minkowski theory into general relativity, the falsehood of the Minkowski formula did not affect relativity; this is because Einstein knew that the logical foundation of his theory has sound, physically sound. Secondly, the time element could not undermine relativity because, ironically, time units are always the same so long as we base all time on the earth-year---and time is used only in units---see APPENDIX ONE below.

claimed that time and space are found to be one and the same thing in a world of 4-D geometry, used time in that research. Without time the same 'time' that the researchers used to produce time from the merger of space and time, is the same time again, as the product of the merger between space and time. There is only one time on earth because everything is based on the earth's orbit of the sun as 'one year', subdivided down to the seconds and so forth. There cannot be time as a result of the merger between time and space! How can they say that it is the same as the time (again) after they have used it to analyse Order and Simultaneity? It does not make sense, however much one may try to say what they mean in clear language. I am trying, in effect, to put into sensible and acceptable words ideas that are basically a tissue of nonsense.

The analysis of Order and Simultaneity, in so far as it is meant to prove the existence of time-producing union of space and time in a 4-D geometry, cannot be regarded as a logically valid investigation to discover the nature of time (and whether or not it is one and the same as space), simply because the researchers had to use time in their analysis of Order and Simultaneity. There is only one time as per the earth-year, and that time was there already---they used it! What they did is 'the application of time'; and you cannot apply a thing to discover itself all over again as if it is something newly discovered. Whether time and space constitute one entity or not, you cannot use time to discover that.

The application of time includes (indeed it begins from), the clock. The clock is our principal instrument for the application of a time system obtained elsewhere—e.g. the breakdown of the earth-year---and deliberately programmed into the clock for reproduction in specific units determined by man. What the clock shows as the reading of time is not an original creation of time, but the reproduction of time---as discovered elsewhere and deliberately programmed into the clock for 'reproduction'. Let me explain this absolutely clearly for the benefit of the reader. The seconds ticked by the clock came from somewhere; somebody decided that they should 'be' of that specific nature. The reason is that the clock can only 'apply' time; it can never produce time. The time consists of the union between the internal time-sense, known as duration, and repetitive external cycles---the years for instance, as we know that the passing of the years is the passing of time. The seconds as produced by the clock are derived from the breakdown of the year---otherwise they could not exist. They are not produced originally by the clock itself. Rather they are repetitions or reproductions of what have been programmed into it. That is "The application of time", or of what has been discovered elsewhere as the specific units necessary to lead up to exactly one year at the end of the earth's orbit of the sun—and start again for another year, ad infinitum. So time proper is the union between the sense of duration and repetitive external cycles. Let us say, for example, that you have a sensation, or are involved in an event, and to know how long it lasted you use some repetitive cycles or regular motions and count the number of the cycles, and say it lasted so many cycles. The earth-year is one such cycle; but it is so long that we have learnt to sub-divide it down to the seconds, and so forth. That is the mechanism used for creating the seconds ticked by the clock. Hence the reading of a clock is "The application" (display, disclosure, demonstration, etc.) of time---

that is, of those units of time obtained from the breakdown of the year and deliberately programmed into it to tick them precisely continuously to coincide with the completed orbit of the sun by the earth. So using the clock for the analysis of Simultaneity is "application" of time. It cannot be true that the union of this application of time and space is what gives us our earth time originally as Minkowski claimed---to the effect that neither can exist on its own. That is tautology.

It is interesting that Minkowski had the audacity to proclaim that because of his theory neither time nor space can exist independently except as produced by the union of the two by mathematics! And as we all know, this view of the world was arrived at by means of the analysis of Order and Simultaneity (originally by Einstein, for, as Russell has pointed out, the space-time concept was already present in special relativity in the idea of 'local time'.) So the time that Minkowski was proclaiming to be incapable of independent existence was nevertheless used as an independent entity by the same Minkowski since you can only use time (as the application of time) to analyse Order and Simultaneity.

To repeat, it was not a valid logical investigation to discover what time is because they had the time in hand already, since they could not have carried out the research without time. What they did was a process (just one process) of applying time or using time as it is. For time and its application are two different things, neither of which can be used to discover the nature of the other.[19] This is my standpoint. Now the analysis and arguments:-

---

[19] The concept of 4-D geometry is so serious that it must not be averred lightly. You need cast-iron proof; for if it is really in existence then it is the greatest mystery in life overall that

Time is one thing; the application of time is quite a different thing, a different operation of the mind. Without doubt many of the mysteries of time are caused through the confusion inherent in the application of time as it essence, in phrases like "Do that in ten minutes". I should think that, properly put, it should be "Do that in ten cycles, assuming that a cycle is one year, or an identifiable unit of time. In our system, however, that is not the case; but we know that ten minutes are parts of the earth cycle for one year. Therefore, theoretically, they are parts of the earth cycle round the sun.

In simple terms, the cycle can be replaced with a tap from a finger. In such a case the command should be "Do that in ten taps of the finger"---a different time system, but time all the same. Since time is always in use (indeed unavoidable), its use becomes something of its permanent nature in the human mind, or the popular imagination; yet the essence of time is different from its applications. The units must be established elsewhere beforehand (by the subdivisions of the year, for instance). They are the essence of time. Creating the units is the essence of creating time, in units, and in units only. The rest is applications of time---of the units of time. Hence time as we know it, in units only, the basic unit is the earth-year. There is no idea more difficult to put into words than this.

Scientists have never recognised this; if they did they would not speak of 'Time Zero', meaning the metaphysical or ultimate beginning of time---of whose time? Under relativity earth time originates from here, and is limited to the earth. This earth time is necessarily discrete and cannot have originated elsewhere from Time Zero to reach today's date by marching through the cosmos, since discrete time implies (or requires) the

---

could never be explained without appeal to divinity. It is ludicrous to use imaginary quantities to prove it.

use of points. Mathematics is necessary, a theory of numbers is essential, and the ability to count the orbits of the sun as years must be there---all of which makes the invention of time human in origin. We must never lose sight of the fact that it is the year we subdivide down to the seconds and even the cesium units (since they have to be related to the second to make temporal sense); yet the year is one discrete unit of time that has to be repeated again and again for time to continue.

Once you accept the year as a valid unit of time, you are logically obliged, also, to accept that earth time is basically discrete, with momentous implications, most of which will face staunch religious objections. On the other hand, in the scientific interpretations of the nature of the external world for purposes of human welfare and so forth, nobody ever makes any allowances for personal, national or international religious sentiments, and that is quite right.

If Einstein is right, and we all believe that he is, then "There are as many times as there are inertial frames, or bodies". In this sense, whose time or what time began from Time Zero?

There may be religious sentiments in this which must be rejected (mathematicians are incurable mystics, and we have to accept that the existence of god is still debated, and seems likely to be debated to the end of the world); but I have always sadly believed, also, that scientists instinctively conceive time as a universal entity, and that earth-time, however it is logically defined, is only a version of cosmic time. Yet Bertrand Russell's interpretation of the Einstein notion of time is that "There is no longer a universal time..."[20] Thus he asked the most

---

[20] He wrote: "There is no longer a universal time which can be applied without ambiguity to any part of the universe; there are only the various 'proper' times of the various bodies in the

important question about human life in the same book, namely: "if cosmic time is abandoned, what is really measured by a clock?"[21] This is on the understanding that Einstein's analysis of Order and Simultaneity proved that cosmic time does not exist, and that time is neither general nor absolute as in the Newtonian sense, but dynamic and changeable---the best description was invented by Lorentz, for he is the real origin of the new notion of time. He called it t^1 or local time, and that it is different from t, or normal time. As a matter of fact, to ask what is really in the clock, when we already have the clock, is an admission that we simply do not know what time really is, therefore the passage of time as indicated by the ticking of the clock is the only evidence we have of the nature of time. This is the gist of my theory of time. To be frank, it is not original; I just looked closely at what the clock does for us as time following relativity and luckily came up with the notion.

I believe scientist still think of time as the Lorentz t, and earth time as local version of t. In this sense the concept of Time Zero is justified, conceivable or logically tenable. But if cosmic time is abandoned, then there can be no such thing as Time Zero. Nevertheless, scientist will reject this suggestion because it is not couched in intricate mathematics. The fact that mathematics often leads

---

universe."---Bertrand Russell, ABC of Relativity, Allen & Unwin, 1925, Ch. V.

[21] If Russell is right, and I personally think he is, then Einstein did not only change or merely revolutionised physics; he practically changed our basic and most important conceptions of the world we live in. It is now seen as a separate world with its own time that bears no relation to any time existing elsewhere in the universe; and which time leads us to theories and suppositions that make our science what it is; and for which reason even the quantum may be just the peculiar product of our unique time, and not existing as we know it throughout the cosmos---a very serious idea indeed.

them astray is usually ignored. The additional idea that there are two aspects of physical theory, consisting of the physical concepts and their mathematical expressions, is also brushed aside contemptuously as evidence of the ignorance of counterintuitive mathematics.[22]

I can assure the reader that I am not affected by such insults from mathematicians and scientists. For what Professor Eddington said is precisely like the Bertrand Russell critique of Minkowski, namely that his concept of 'interval' lacks non-technical meaning; yet such a meaning ought to exist if interval is as crucial as it is in relativity. Well, for me, to be completely impartial, the Minkowski theory is not crucial in relativity but only in the mathematical interpretation of it as proposed by Hermann Minkowski and which is described as arbitrary and fictitious!

The point is either Einstein is right or he is not. If he is right, then all those theories of Time Zero, curved space time, and time travel now making the rounds of learned journals are not only misleading but are positively distorting relativity and physical theory as a whole. I am not concerned for myself personally because, although I cannot command mathematics to the level of the professionals, I am satisfied that both Bertrand Russell (a great mathematician, logician and philosopher of genius), Professor Sir Arthur Eddington, a great mathematician and astronomer, and Professor

---

[22] Yet in the Preface to his book, *Space, Time and Gravitation* (Cambridge, originally published in 1920), Professor Eddington made the same point. He wrote "...the mind is not content to leave scientific Truth in a dry husk of mathematical symbols, and demands that it shall be alloyed with familiar images. The mathematician, who handles x so lightly, may fairly be asked to state, not indeed the inscrutable meaning of x in nature, but the meaning which x conveys to him."

Whitehead---mathematician, logician and philosopher---all agree with Einstein.[23] So all one has to do (which is what I have attempted in this little book) is to ensure that one's suppositions accord with the views of these great men of science and mathematics. And, specifically, concerning the Einstein theory of time, I give a quotation of what Professor Eddington wrote and I advise the reader to take particular notice of the very strong words used by a noble scholar as great as Eddington: "Prior to Einstein's researches no doubt was entertained that there existed a 'true even-flowing time'[24] which was unique and universal [after this introduction Eddington takes up the question of time and true time, before returning to the defence of Einstein. As I have indicated above, I believe it is instinctively felt by all scientists---including Professor Eddington on this evidence---that there is time and there is earth time][25]...Those who still

---

[23] It would be most unwise to regard these giants of 20th century thought as in any way old-fashioned or that their views are out of date because science has moved on. Actually I think we ignore them at our peril, for we have not yet even began to study their ideas properly, particularly with regard to their interpretations of relativity, which is still not well understood, as one editor put it way back in 1988: "The theory of relativity is eighty years old but we have not got used to it. In three-quarters of a century it has not succeeded in changing the habits of our thoughts..."--- not at all, partly due to the distortions brought in by Hermann Minkowski, I think. (See COMMENTARY ON THE THEORY OF RELATIVITY FROM *The World of Mathematics*, Published by Tempus Books of Microsoft Press, 1988, Vol. 2, p. 1083.)

[24] Since then we have actually had only 'discrete time', a cornerstone of my suppositions about time.

[25] They are wrong, of course. Earth time is discrete, and that is what matters. Nobody knows of any other time existing elsewhere that has any effects on man or his earth time. Einstein's demolition of cosmic time was wrongly interpreted at the time. We know better now.

insist on the existence of a unique 'true time' generally rely on the possibility that the resources of experiment are not yet exhausted and that some day a discriminating test may be found. But the off-chance that a future generation may discover significance in our utterances is scarcely an excuse for making meaningless noises."[26]

In spite of this strong condemnation of any time system opposed to the Einstein notion of time, Professor Eddington supported the use of the Minkowski theory that he himself described as 'fictitious and arbitrary'. And probably because he sensed the discrepancy in the situation, he did not even once mention the name of Hermann Minkowski in this important book which is wholly devoted to the mathematical interpretations of relativity, except in the Bibliography.[27] The reader must

---

[26] *The Mathematical Theory of Relativity*, by Professor Sir Arthur Eddington, Cambridge, 1930, Ch.1.1.)

[27] All through the book he treated the Minkowski idea as if it is textbook knowledge and part of relativity. Now, it is obvious that it is what mathematicians hoped 'is' the true state of nature---that is that space is 4-Dimentional incorporating time, so that time does not have to be added to phenomena as in the 3+1 formula. Yet this same great scholar and mathematician has warned us that we must never forget that the Minkowski formula is arbitrary and fictitious, as he put it: "...his space and time reckonings are imaginary surfaces drawn in the world like the lines of latitude and longitude drawn on the earth. They do not follow the natural lines of structure of the world, any more than the meridians follow the lines of geological structure of the earth. Such a mesh-system is of great utility and convenience in describing phenomena, and we shall continue to employ it; but we must endeavour not to lose sight of its fictitious and arbitrary nature." *The Mathematical Theory of Relativity*, Ch.1.1. Now, who is he talking about---Einstein? Of course not. This is a perfect description of the basic elements in Minkowski's 4-D geometry

not blame me if I make too much of this anomalous situation. For, as far as I am concerned, whether Minkowski was mentioned in name or not, I know that all writers on relativity accept his 4-D-geometry, or the four-dimensional interpretation of physical reality, instead of the original Einstein formula known as 3+1.

For the reality of the matter is that, since Einstein and Minkowski, the whole of theoretical physics, astronomy and cosmology (which last subject is for the interpretation of the cosmos scientifically or logically), have been built, or reconstructed on the Minkowski 4-D geometry proposal, being the idea that space and time are unified into one entity (therefore $s=ct$), expressed with the ubiquitous term 'space-time' (the same theory Professor Eddington has described as arbitrary and fictitious.)

Altogether Minkowski appears to have been very lucky, perhaps the luckiest theorist in the history of science, for even the very logical and fiercely secular philosopher, Bertrand Russell (acknowledged as the founder of the philosophy of science), also used the term space-time all through his book, *The Analysis of Matter*, while condemning the mathematical formula (or equa-tion) upon which it is based as arbitrary and an artefact compounded for the convenience of the mathematician---note that he does not say it is compounded for the rational or scientific study of nature; neither does he say it is the true nature of physical reality, no. Russell wouldn't make that mistake because he also described the theory as based on arbitrary assumptions. It has rather been (perhaps specially) created with questionable mathematics for the benefit of mathematicians to, in my view, 'play' with; yet it is now promoted as the essence of physical reality so that the whole of cosmology is based on it. I quote the Russell statement below: "SPACE TIME, (sic) as it appears in mathematical physics, is obviously an artefact, i.e. a structure in which materials found in the world are compounded in such a manner as to be convenient for the mathematician."[28]

Scientist always stress the mathematics of Minkowski, calling it detailed and complex, as if these are the essential merits of a theses, when in fact clarity and simplicity, as in the Maxwell equations, or the Einstein $E=MC^2$, is rather praiseworthy.

As we have seen, even Einstein praised the Minkowski mathema-tics. Yet mathematics is not the ultimate arbiter of truth in physical theory---what is proposed in mathematics must be physically discovered or the proposal is null and void, and can never become part of physics proper.[29] Even then the space-time mathematics seems to have been specially created (and mathematicians are allowed to do that), for the purpose of merging space with time. Some aspects of the formula are logically untenable. For example, he relies completely on i to represent imaginary time co-ordinates, and I quote from Einstein himself to escape the insults of mathematicians: "...we must replace the usual time co-ordinate t by an imaginary magnitude..." (RELATIVITY, Part I, section 17.) So time is represented with i, an imaginary magnitude, or quantity; yet there is no such thing as imaginary time. We have enough difficulties trying to define the time we have, let alone

---

[28] *The Analysis of Matter*, Ch. XXXVI.

[29] Take for instance, the fact that the whole of mathematical physics is still searching for the Higgs boson; and if we can believe the press, billions of dollars have been spent creating the 'Large Collider' to detect it, and unless it is found, the whole complex mathematics used for predicting it will mean nothing, and the money is lost. The same thing, as I have said, happened to the eather for the propagation of light---what happened to the numerous mathematical proposals about its existence?

imaginary time! But, still, I have to stress that this theory of 4-D geometry is now dominating all physics, astronomy and cosmology. Astrono-mers in particular are always referring to "The dawn of time" in reference to events in the cosmos. But the dawn of whose time, if Einstein is right to say every frame has to have its own time? The fact of the matter is this: it is not just that Einstein said so; it was Lorentz (not Einstein) who discovered that time can begin from anywhere.

Furthermore, by the analysis of Order and Simultaneity Einstein confirmed this idea. The corollary is that, in summary, (1) "There are as many times as there are inertial frames"; (2) there is no longer a universal time; (3) there is no absolute time, and so one second here is not the same everywhere else; (4) time is dynamic and variable from place to place; (5) it originated from this planet and limited to it; (6) we don't know of any other time anywhere else in the universe, and using earth time to apply to other regions of the universe is logically untenable; (7) and that, as a result, earth time is automatically discrete time---which is serious because discrete time has momentous implications most of which are strange and unfamiliar and man is going to have to get used to them to make the world more rational in conception than it is at present. Hopefully, politics will follow suit.

Ironically, these scientists take time *as it is*. That means the time is in existence already before being merged with space. *It does not mean the time comes to exist as a result of the merger*. That would not be logically defensible because the time is what was (or usually 'is') used to analyse Order and Simultaneity in the first place. Let me stress this. It does not imply that we get the time as a result of its union with space---that is tautology, because the time was in existence already before being used in the analysis of Order and Simultaneity, purporting to show that space and time are unified into one entity. Yet that analysis did not conjure up (and could in no way have conjured up) the time into existence as the product of the "merger between space and time". For the time was there already; and it is what was used to analyse Order and Simultaneity. It did not come up as a result of its being merged with space---that is tautological and logically untenable. In very serious fields such as theoretical physics and technical philosophy it is an insult to our intelligence to say this; yet Minkowski not only got away with it, he actually became famous as the man who made relativity accessible to mathematicians, which can only imply that relativity is not yet properly understood.

I conclude, therefore, that time cannot be in space as a result of mathematics that used it (as it already was) to prove that it is part of space---as analysed with the time in hand as aid! This is wrong, surely? But then some of the giants of 20th century thought (Russell, Eddington, Whitehead), have condemned it as arbitrary; even so it dominates theoretical physics and cosmology still.

The reason is that time and the application of time have been conflated; otherwise the researchers could have noticed that using the time in the analysis is 'application of time'---which implied that the time was in existence already otherwise the analysis of Simultaneity, which requires time, could not have been carried out. It is logically impossible to use an entity you already have in hand (as the application of time) to created that same entity (as the 4-D geometry was meant to create time), and then conclude that that same entity could not have come into existence---for that is precisely how Minkowski put it in his Cologne lecture---without its being conceived as the product of the merger between that same entity already in existence and something! This is worse than children magic in a pantomime. From this it would have seem illogical for anybody to say he has invented mathematics for equating space to time---yet that is what mathematicians are claiming with what amounts to religious fervour. That it leads, inevitably, to a distortion of physical theory does not seem to be part of anybody's concern. Rather, they are all waiting for another revolution from somebody like Einstein.

It seems to me quite obvious that the dichotomy between time and the application of time (as two distinct operations of the mind) has been overlooked in science: the mathematics for the analysis had to use time as it was already. So how can that same analysis point to the time being rather discovered by the very medium that the researchers were using time to analyse? It does not make sense. It arose out of the confusion between time and the application of time.

Too often the interpretation of the nature of time is mistakenly derived from the application of it. Again let me emphasise this: time and how it is applied are two different things. Unfortunately, we can only illustrate time with how it is used, because how it is used consists of how it is passing; and unluckily for mankind how it is passing is what it is! We only know of time by the way it passes. All we can ever know of time is how it passes by.

It gives us the wrong notion of time; but there is nothing we can do, for no one can define time logically. We can only know time from how it passes. The unit is the time; the metaphysical nature of time is its unit---the year is one unit of time; and we know it through how it passes; and it is always passing. It never stands still. When it returns to the median point, the year is spent---and does no longer exist; we have to start another year. When the unit of a discrete time system is expended, the time is no more: at the stroke of midnight, it's gone, and we start counting from zero to another unit of time.

Thus we can only know time from how it passes, and that is usually taken from how it is used. When time is passing it is being used. It is very important to realise this. When time is passing we know it from how it is used; yet the passage of time is its metaphysical essence: we know time from how it passes, unit by unit: for instance, year after year after year---or second, second, second, on and on. We know only of the passage of time not its real nature. That is what we have created as time. The year, again, is just the physical orbit of the sun. Yet the passage of the year is the passage of time as far as we (or mankind) are concerned, even though it is a physical activity. It is man who regards it as 'one year', and a unit of time. We are using repetitive cycles in nature as time. So we only know the cycles and how they pass "as time". Since all other units of time are derived from the earth-year, they are also parts (fractions) of the cycles we regard as time, or use as time.

Again, the passage of time is always known through how we use time, simply because time is always in use; it can never be suspended.[30] The passage of the year is the passage of time, yet the year is always in use as it passes, for its passage is its essence. To separate time from how it is used is all but impossible except in philosophy. But people are not professional philosophers. They just use time. Even scientists conflate time with its applications---hence the notion that time can be used for the analysis of simultaneity as 'evidence' that time is the same thing as space. I agree that there can be some mathematics that shows that time seems to be merged with space, but to propose by the use of the same time in the analysis of simultaneity to conclude that time is the same as space in a world of 4-D geometry is obviously illogical, the worst kind of tautology.

---

30 I am deliberately repeating this for the purpose of planting it firmly in the reader's mind irrevocably, since we are all going to have to get used to it.

Earth time is necessarily discrete because it is entirely based on the year, and the year is just one unit of time that has to be repeated to continue. But if time is discrete then we can only know time from its constituent units, which must be recognised as beginning not from the second but from the year, simply because the second and all other units of time are derived from the year as fractions thereof. In this sense, time can be defined as "The units of space traversed by the earth round the sun". The problem of ultimate definition arises from how we see the year. It is utterly impossible to define the year (as our basic unit of time) in logic.

But relying on the year for all our time (units) makes earth time discrete. Discrete time has momentous implications. For a start, it means the concept of Time Zero is untenable, since discrete time cannot march all through the cosmos to reach today's date. Curved space-time is also false; space may curve, but discrete time is independent of space; therefore when space curves it cannot take time with it---discrete time is spent when the unit of it is no more, that is why the year is ended at the meridian point, and we have to start counting another year second by second after 31st December. It cannot curve because it is not conceived as a thread, but rather as consisting of discrete units---the earth-year for instance. If there is no such thing as curved space-time, then, of course, time travel is a nullity, if not complete nonsense.

According to Bertrand Russell, "The most remarkable feature of the theory of relativity, from a philosopher's standpoint, was already present in the special theory: I mean the merging of space and time into space-time."[31] This is correct. It makes the Minkowski fiction even redundant. For I have always believed that it was the Lorentz local time concept that Einstein developed

---

31 *The Analysis of Matter*, Ch. V.

into his theory of frames, and the concept of dynamic space-time; in any case, we have it on the authority of Lorentz that the local time notion led directly to special relativity, and that he didn't discover the theory because he failed to attach due importance to the local time idea.

The local time idea implies that "There are as many times as there are frames". In other words, time can begin from thousands of places, or localities. How? Simply by the application of points to space to create time units, or time cycles---as we use the earth's orbits and the Meridian point to create our basic unit of time out of which all other units, including even atomic time units based on the second, are derived with points.

For, after all, I repeat again, time can never be logically defined. The nearest we can get to knowing the nature of time is by using repetitive or regular cycles we call 'years', and further sub-divide the year into the seconds and so forth. Russell is right because the concept of space-time (as meaning that time is derived from space, not from 4-D geometry) is clearly implied in the Lorentz local time idea which Einstein adapted. And it is right to call it, too, "space-time", as it is derived from space with points.

Once Lorentz was able to start a time system different from normal time, intellectually man was liberated from the tyranny of rigid absolute time, generally believed to be divine and immutable. The changes that this secular time brought to everything we do in life (science, philosophy, religion, social theory, et al) are so serious that the nature of time will continue to be debated till the end of life on this planet. For human existence consists entirely of two natural elements: sentience (or the life), and the time for its rational regulation without which living the life rationally would not be possible. As such the importance of time is second only to the existence of the life itself.

Einstein correctly said the Lorentz local time is "Time, pure, and simple". And Lorentz agreed, saying he failed to discover special relativity because he did not take the local time notion serious. Einstein did, and as they say, the rest is history.[32]

---

[32] He said, "The chief cause of my failure [to discover special relativity] was my clinging to the idea that my local time t^1 must be regarded as no more than an auxiliary mathematical quantity." Quoted from Abraham Pais's Biography of Einstein, '*Subtle is the Lord…*' Ch. 7, Oxford, 1982.

SUMMARY AND CONCLUSIONS

Once the four-dimensional space (including time) is dismissed as flawed for being based on imaginary time coordinates, we must conceive time as the product of points as applied to space; the points are important for otherwise we could not have time in units. Being the product of points and space, we can still call it 'space-time', or properly, 'space-timed'.

That does not mean the whole of secular time is external. As Professor Eddington has pointed out (in my support, I am glad to emphasise!), man has his own personal time-sense known as duration. This we sub-divide with external cycles. Without making the narrative too technical, it must be realised that every unit of time is derived from the yearly cycle, therefore it is a fraction of the cycle---so every unit of time is a cycle of sorts. Thus if you encounter an event, and it goes away, say a period of extreme bad weather that you remember pretty well, how do you tell how long it lasted? If the yearly cycle was a short one, you cold just mention the number of cycles the bad weather lasted. But since the year is very long, we have sub-divided it down to the seconds; so you will mention the appropriate unit of time, being one of the sub-divisions of the year. In this way we apply external cycles to events as time. Time is both the internal time-sense and the external cycles taken together. This is the reason I define time like this: Being is time without mathematics---so you need the 'Being', you need the life's existence.[33] What we call

---

33 I must concede that 'Being' looks suspiciously like time, but then we have to have time in units---the basic time, the earth-year, itself only runs one by one. This is the real mystery (or irony) about time---why is it entirely based on the year, which is just one unit of time? Anyway, the best way to achieve the individuation of time is to use external repetitive cycles and count them as the rates (or units) of time, such as we use the year for.

time and which we mechanise in the clock is 'Being' as subjected to (or sub-divided by) external cycles, or external mathematical cycles---poetically shortened to the phrase: "Being subjected to mathematics".

The next quandary is how time passes, and why it is continuous, taken as one question. This is known in religious circles as 'The Problem of Perpetual Time'. For the religions this is a question that man can never answer. For one thing, it is stressed that time is there as we are born, and there as we die---so it is assumed that it is perpetually there as something of an eternal mystery best explained with the believe in God's existence. Even if this is true, and I am certain that it is not, there is no obligation to prostrate ourselves, noble human bodies, in homage to God, none whatsoever. Besides, it is our fellow (mostly cunning and tricky, wicked and corrupt, selfish and greedy man) who tells us how and who to worship in a completely arbitrary manner. Let us face it, if God comes down and orders me to worship him, I will ask him why. Yet we don't do that in life. Largely corrupt and wicked haters of mankind (or slave owners) collude with the holders of power to compel man to worship; that is how it began, even if it now appears to have become second-nature to the gullible. A writer on time is necessarily compelled to comment on religion to dispel the numerous religious false ideas about time.

The passage of time becomes easy to explain when time is understood as the discrete product of points as applied to space. As stated above, discrete time, like the year, can only pass by and seem continuous through the succession (or procession) of its units. Who decide that this will go on forever? You may ask. The answer is 'nobody'; impersonal mechanics in nature determine that; but it is not going to go on forever. Astronomers can tell you exactly when the sun will dim, bringing all life to an end on this planet. When that happens, and there is nobody to count the orbits of the sun as 'years', earth time will end abruptly. It will end, also, if the earth fails to maintain its orbits of the sun for any other reason.

The important point, here and now, is that thanks to Albert Einstein, we now have a credible answer for the passage and continuity of time without the need for the mythical arrows of time; we can also fairly accurately define the nature of time in so far as the passage of the years is the passage of time. So we can say we know time from how it is passing. Above all, I implore the reader not to lose any sleep about time travel; by discrete time as we have on earth, it is absolutely impossible to realise time travel as a fact rather than a mere dream. There appears to be no sort of hidden meanings in time to frighten us any more. And we know now how it passes and seem continuous as a matter of valid logical deduction rather than speculation about the mythical arrow of time.

APPENDIX I:

Time and quantified time

We are all fond of using the word 'time' loosely to refer to the passage of existence in any form whatsoever. That may be called 'the unscientific' notion of time. In logic, science and philosophy, however, time is what Professor Richard Feynman called 'how long we wait'. This translates into the concept of 'how much time', or quantified time, so as to be able to tell how long we wait in mathematical language for universal application.

In a serious discussion of time, it does not make sense to just mention time. The context of any proposition (in science, mathematics and philosophy) must always show or imply the sense of 'how much time' in it, or expressly show the quantity of time proposed. Of course, time may pass when one is not conscious of it. But in all cases, when one wants to know how much time has passed, or will pass (as in futuristic propositions), mathematics must be used to quantify the time. And let me stress again that we quantify time by the use of external cycles in union with any sense of duration of anything whatsoever.

Quantified time is 'time in a clock', any clock at all. And the clock, any clock, can only show time as independent of space. Space-time is automatically quantified as it is derived from space with points, which is the only reason for calling it 'space-time'. Discrete time can only pass through the succession of the individual units. On this point, Leibniz was absolutely right when he said time is succession. What was lacking in his day was the concept of discrete time; with this new concept in our post-relativity world, we can now see clearly as to how time passes and seem continuous through the succession of its separate and individual units: second, second, second; or minute after minute after minute. Plus the hours, weeks and months all the way to the year, which also passes in the form of year after year after year.

It may seem surprising, the springs of a thousand legends, giving rise to supernatural speculations, that we have an extremely ingenuously smooth time system, so cleverly structured that it is there when we are born and there as we die, and always passing by. For this reason we know that "Time does not wait for anybody", not even Kings and Queens and Presidents. Even surrendering one's Kingdom and all possessions for a moment of time cannot save the most powerful Queen on earth. Scrutinised under a logical gaze, however, time is not so rosy; it is only one moment, repeated to pass by and seem continuous so that arithmetic can be applied to its accumulations. [34] This, as we know well, happens when we reckon time for futuristic planning, and backwards as history.

But for the union between the sense of duration and external cycles giving us units of time out of the moments of time, time for the clock would not exist at

---

[34] Let me explain that space-time is necessarily discrete. We have only recently come to understand space-time from Albert Einstein; yet time has always been discrete, consisting of only one unit (or moment) of time---of whatever length. For there is only one year, and all other units are obtained from the year in the form of separate units of time. To get more years we repeat the one year exactly. Thus we have second, second, second; or minute, minute and hours and so forth. Each is a moment (or a unit) of time in its own right. The notion is best illustrated with mechanical devices: if you set a timer to regulate the working of any mechanical device, when the time ends, the machine will stop because the time allowed has ended---time, even in this sense, has a 'beginning and end' based on human activity, as Lorentz discovered, and Einstein was right to extend the notion to all time, or time *per se*, exactly as he put it, 'time, pure and simple'.

all. Presently philosophers see time as rather a straightforward pragmatic entity, albeit not as simple as it is normally supposed. It is partly a confidence trick, which makes the clock work continuously, the trick of continuity is in the repetitions of the seconds, or of the units of time, all of which are to be understood as single moments---which are the realities--- of quantified time. It is also partly physical (using physical cycles for the process of quantification); and partly philosophical, i.e. according to Einstein without time physical reality is indecipherable, or cannot be properly (accurately) determined.

What Is Measured By The Clock?

Our time is based on the repetitive orbits of the sun by the earth, and evidently the earth never stands still. If ever it does stop going round the sun, our time system will be completely nullified; but, of course, life will go on. It is inconceivable that all life will be extinguished instantly the moment our time is (mathematically) nullified in the sense that only quantified time would be lost. This is the best proof there is that life is not based on "time allowed", as the religions believe; rather time is a union between the sense of duration and external cycles---therefore man had something to do with the time we have in the clock, the only reliable time, as quantified time.

All the religions speak of "time allowed" for the duration of a man's life. They had to, because the nature of time is easier to explain as a providential bounty than anything else. To be honest, without a cosmic explanation for time, what is time, or, to put the question in another form, what is the origin and essential nature of time? Of course it is assumed that the clock measures time---but from where? And what is it that the clock measures?[35] The clock maker will say he invented the

---

35 Without the explanation that what the clock measures are cycles of duration, or duration reduced (or converted) to repetitive cycles, metaphysically interpreted as a union between duration and its conversion to external cycles, time can never be logically accounted for. We would just go on using it---but in what form? In the form of units (year after year after year, and all the seconds and so forth derived from the year); yet that means the same thing, namely, a union between duration and its conversion to external cycles. For the year is only a cycle of the sun. It is not time. It is the practice of humankind to call it 'a year; and we use it as our basic unit of time, as a matter of convenience. Otherwise in nature it is not time. As a matter of fact, we can use something else---we can tap the finger, for instance.

clock to reckon time in the sense that everybody knows---but what is that sense of time?

When it is postulated that general time permeating the whole cosmos (and therefore the same everywhere) does not exist, the first implication is that every body has to have its own time; it is not coming from the cosmos therefore it must have originated on this planet. So let's find out how it all began. That is the first implication. The second is that, as a result, cosmic time is abolished---although it sounds tautological, it still has to be emphasised, as well, and most clearly because the 'cosmic time instinct' is permanently ingrained in the human mind. One reason is that time cannot be suspended; but the more cogent reason is sheer intellectual incompetence plus fear of the unknown. We are always using it, and so it does not make sense to just insist that it is not there.

If it is there, and did not come from the cosmos, how did it begin? And the obvious fact is that it is always there. Even before we are born, and also as we die to leave it behind. Yet it cannot be supposed that each body's time is a version of something 'naturally existing', whether it permeates the whole cosmos or not, with the necessary but illogical (little 'academic') proviso that it may not be the same everywhere but varies with individual bodies in accordance with unknown natural laws.

It is plainly evident that this erroneous sense of time dominates scientific thought. Hence time is not defined in physics; and as a result, the Minkowski fiction makes sense to some scientists, including even Albert Einstein himself. So far only Professor Arthur Eddington has redeemed physics by warning that it must never be forgotten that the Minkowski formula is "fictitious and arbitrary"---but they have chosen to ignore him.

Thus Russell's query is important, namely, "If cosmic time is abandoned, what is really measured by a clock...?"[36] My answer, of course, is that outside the union between the sense of duration and its conversion to repetitive external cycles, time does not exist to be measured. The very act of 'measuring' is the time in essence---like moving from point to another point, time is going, so that time becomes 'relation between points'. The cycles are time units (the years, for instance), and the time units constitute the time: a year is a cycle, but it is our time, the basic unit out of which all other units are derived.

However, the cycles are the creation of man for the sole purpose of converting the sense of duration (of any thing or any event, like the period it will take to reach the village from the farm before nightfall to avoid predators), to his time units to guide his activities. So the clock does not measure time; it rather reproduces units of time programmed into it repetitively---second, second, second, and so forth. It should be remembered that the seconds are put there by the clockmaker; but where do they come from? The answer is that they come from the subdivisions of the year. Otherwise the time does not exist anywhere to be measured---the units constitute the time. Without the year there will be no seconds, and the like, all of which are derived as subdivisions of the year. As hinted above, you can even dispense with the year and its subdivisions and tap your finger, if you will not get tired. A million taps means it is time to go to bed, and so forth; outside the units of time, time does not exist to be measured; but the units are the creations of man as quantified time to record the passage of existence in his experience in manageable units for cultural purposes.

---

[36] ABC of Relativity, Ch. 4.

APPENDIX II:

The principle of mathematical equivalence

In nature there is reality and our perception of it. In the word 'perception' everything man does in life is implied, including mathematics, since we can only act by perceiving the true nature of the physical world; I am using the word in a sense akin to 'experience'. The problem is, mathematicians normally are permitted to imagine things to satisfy their nostrums, so that they do not rely solely on their percept; however outrageous, they can defy reality, logic and common sense, and leave it to the applied mathematicians (physicists, astronomers and cosmologists) to find out whether what they have assumed is there in nature, so that their theories based on it can be seen as true or not. In no other profession is this sort of thing allowed. Even one of the greatest mathematicians Britain has ever produced, Professor Sir Arthur Eddington, criticised that common mathematical tendency in his book, The Mathematical Theory of Relativity. I have quoted him above in the text, but it will do no harm to repeat it as it is vitally relevant here. He said: "The pure mathematician deals with ideal quantities defined as having the properties which he deliberately assigns to them. But in an experimental science we have to discover properties not to assign them..." The principle of mathe-matical equivalence should make them think of the practical consequences of their imaginary properties, although I doubt it, but that is another matter. The rule is that mathematics should not seek to make the basic features of nature what they are not quantitatively; any such propositions are bound to falter. Note that we are talking only of basic phenomena. By the very nature of man, it seems he can make qualitative changes in peripheral nature not quantitative changes in the fundamental aspects of nature, and time is the second most fundamental feature of both nature and life.

The principle means that, in effect, one cannot use mathematics (sometimes defying comprehension) to state, say, that there are ten trees in a field, and propound theories about them if, in actual fact, there are only two. This is slightly different from assigning imaginary properties to nature. It is different because it relates to 'quantities'. Six into four won't go, or something like that. The principle of mathematical equivalence rules that, to accord with physical reality, one can only talk about two trees, or as things are not as the mathematicians want them to be. Nature is not there for the convenience of mathematicians; it is neutral. That was the advantage we gained when the ancient teleological interpretations of phenomena was discredited. Therefore this rule is not to be scoffed at. I regard it as one of the strictest doctrines in logic and metaphysics.

It is not often realised how progressive is the study of philosophy. Quietly but surely, many entrenched myths from our primitive past are being discredited one by one by philosophers. One of them is teleological argument. With that and many other ludicrous intellectual fashions out of the way, it is unacceptable to regard any concept as 'compounded for the convenience of the mathematician', as Russell defined the Minkowski theory of space-time. Some day, we may get scholars writing about the many myths philosophers have discredited through their quiet researches to foster science and progress generally. So I regard this principle of mathematical equivalence as a strict and necessary doctrine to prevent mathematicians arrogating the power and right to alter nature quantitatively in the fundamentals of physical reality. We shall, and should, continue to alter nature qualitatively to our benefit---gardens, buildings, roads, cities, waterways, canals, railways, bridges, tunnels, all science (bar destructive devices), and all art, sports and so forth. They do not change nature but beautify it; but quantitatively, never.

We cannot make one object two, or two objects one, physically. It is not possible realistically. Not in reality only in the imagination. The only one I know of that has achieved any kind of academic adherence in the strictly rational post-relativity era is the Minkowski formula, but then it is regarded as fictitious. Thus mathematicians who rely on it must know that they are falsifying their nostrums.

The origin of the rule will help the reader to understand it well when spelt out: it occurred to me when I was pondering Hermann Minkowski's claim to have made time and space into one entity as from the moment he outlined his theory, as previously quoted, in the following outrageous (even cheeky) statement: "The views of space and time which I wish to lay before you have sprung from the soil of experimental physics, and therein lies their strength. They are radical. Henceforth [that is, from the moment of his lecture] space by itself, and time by itself, are doomed to fade away into mere shadows, and only a kind of union of the two will preserve an independent reality". This is to make two things in nature into one with mathematics ('a kind of union of the two…'). So he knew they were two independent things. How could he have made them one from the very moment of his lecture? He spoke of experimental physics. In fact, the only experimental evidence pointed to time being 'local' in nature; and Einstein adopted it in his special relativity. There was no suggestion that time had been found to be inextricably intertwined with space---rather the suggestion was that time could not be had without space; and that once you have space, you can create your own local time. What Einstein did was to interpret local time to mean "The only Time" we can have.

The actual physical reality known to be in existence was precisely as Minkowski himself stated it---namely, that time and space were two separate things. But it is interesting that he sought refuge in experimental physics. In that sense he did not breach the principle of mathematical equivalence. It shows that he was really a very good thinker; he had to be that good to convince Einstein to adopt his formula for general relativity, which came ten years later. The unfortunate thing for him is that the evidence he cited was really irrelevant to the claim he was making. He needed physical support that time and space are inextricably intertwined and therefore constitute one entity. The evidence that had been discovered by Lorentz and Einstein was that time was essentially local in nature, leading to the supposition that 'there are as many times as there are bodies', and that, additionally, time is different in different places, and also under different conditions. The principle of mathematical equivalence can be used to refute Minkowski's claim to have made them into one entity as from the moment of his lecture.

The rule stipulates that he could only have spoken about time and space as they actually were in physical reality, which, as he has admitted, were two separate entities. The reality before Minkowski was that there was space, and there was time. Even the great Einstein himself made them independent in his special theory of relativity. So it did not surprise me that Professor Sir Arthur Eddington and Bertrand Russell described the Minkowski proposal as arbitrary and fictitious. However, it did surprise me that mathematicians ignored this strong condemnation to claim that they could not understand Einstein's ideas without the Minkowski fiction.

This made me sit up and think, think of a principle to require mathematicians to relate their suppositions to exactly the nature of physical reality laid out before them, not as they would wish it to be to accord with their nostrums. The result is that I came to the conclusion that mathematics can only mirror reality, not to alter it with mathematics alone. So the principle of mathematical equivalence is this: Mathematical statements (or equations) must strictly accord with physical reality. It means no mathematical quantity can exceed or reduce what the actual physical quantity is. No mathematics can make one thing two, or two things one, without physical divisions and unions. As applied to Minkowski, he failed because, as Professor A.N. Whitehead has pointed out, time and space still pass through nature as two entities, not one. He could not achieve the physical union---it was, alas, only mental, imaginary.

APPENDIX III:

Why space on its own is not "space-time"

In Einstein's special theory of relativity, we learn that, "In the absence of gravity, space and time are distinct entities. In the metric of special relativity they play distinctive roles."[37] Nothing in special relativity has changed since then to make all space "space-time". Yet in all their suppositions cosmologists and astronomers always refer to space as space-time.

Let me set out the facts as they are at present, as argued all through this book, and hope they will see the light. To begin from the proper beginning, the whole idea of space-time comes from H.A. Lorentz; until then space was space and time was time. It is true that in special relativity Einstein made space and time dynamic rather than the Newtonian absolute; but being dynamic merely means they are changeable under different conditions. But about time alone Einstein avers that he was able to complete special theory of relativity five weeks after he gained the insight that the Lorentz idea of 'local time' can be defined as 'time, pure and simple'. So let us examine the Lorentz notion of local time.

H.A. Lorentz found that time runs slower when in motion, known as "the dilation of time as a measure of moving clocks". He could not understand why and literally put it aside. He called it 'local time' or t^1. To him it was not 'the true time' but a mathematical auxiliary or curiosity---not very important. Time, he said, was time, denoted with t, and t^1 was something you get as your local time, but certainly not applicable in the outside world as time, because it was a mere mathematical curiosity. May I remind the reader that all this has been given in detail in

---

[37] Professor Jeremy Bernstein, in *ALBERT EINSTEIN: and The Frontiers of Physics*, Op. Cit. p110

the text above? I have even mentioned Lorentz's own statement that he thought he failed to discover special relativity because he did not regard time dilation as of any importance.

Strangely, however, as one of his brain waves, Einstein worked this into his theory of frames. The dilated time was 'local time' indeed---the time of your locality. Now, if the universe was fragmented, then local time would be somebody's time, which to him would be running normally like any other time, but to outsiders, would be running erratically (or slowly, in this case.) In actual fact, that was the case with the Lorentz discovery. People outside the moving clock would see it as running slowly; but those carrying it in the moving vehicle would notice no difference in its performance. That is the genesis of the Einstein theory of frames, I think. Otherwise time was separate from space. What you will find is that it varies under different conditions, simply because everybody has to have his own 'local time' in his locality or inertial frame. But since time is continuous, and having made it a separate co-ordinate in the study of phenomena, dynamic space would have different time co-ordinates at every turn. We recall that Bertrand Russell has stated that from the sun's point of view the tram never repeats a former journey---because the time co-ordinates would be different.[38] Since time is a separate co-ordinate in the determination of physical reality, a different time co-ordinate implies a different situation, different physical reality.

This was the state of affairs when Hermann Minkowski came in with his theory of 4-D geometry making time part and parcel of space---all space. So that

38 He wrote, "We think of a tram as performing the same journey every day, because we think of the earth as fixed; but from the sun's point of view, the tram never repeats a former journey..." (From The Analysis of Matter, Ch. VI.)

cosmologists and astronomers sometimes call his theory "The Minkowski Space", or "The Minkowski Universe", meaning that all nature is subject to the 4-D geometry, where time and space constitute one entity. But let us swiftly add that the foremost mathematical interpreter of relativity was our own Professor Sir Arthur Eddington, the man who confirmed the general theory of relativity. He wrote the definitive book on relativity, called The Mathematical Theory of Relativity. About the Minkowski 4-D Geometry, he stated clearly on Page 9 (Ch. 1.1.), as already quoted, "*Such a mesh-system is of great utility and convenience in describing phenomena, and we shall continue to employ it; but we must endeavour not to lose sight of its fictitious and arbitrary nature*."[39] He was not the only great mathematician who described the Minkowski formula as arbitrary. Bertrand Russell also said it was based on arbitrary assumption. As quoted in the book, he made it plain that because of that the derivation of the Minkowski 'interval' as time from space was illogical, or, in effect, invalid.

Let me try and explain again the reason mathematicians still adore the Minkowski theory---even though they know that it is fictitious: it makes things easy for them. Yet it is not true. They accept the novel Einstein notion that time must be made a distinct co-ordinate in the description of phenomena. You see, the problem is that Einstein made all time (any sort of time) 'local time'-

39 The emphasis is mine. I have had to mention this several times, because, quite honestly, I am outraged by the mathematicians' desire to perpetuate the Minkowski formula as if it is really true of physical reality---yet it is not, and they know it. At least one of their own numbers told them so. It is true that Professor Eddington did not mention Minkowski by name, but, obviously, he wasn't referring to the Einstein theory, calling it fictitious, a hanging offence!

--the time you create for your own local purposes, as Lorentz had discovered. Einstein extended the Lorentz idea to all nature. Every time thenceforth became somebody's local time. But with the universe being fragmented, it was impossible that one system of 'dynamic time' (as opposed to 'absolute time'), *could apply with equal validity to all fragments of the universe*. As a result he said there are as many times as there are bodies in the universe. Nobody can contradict Einstein on this matter. But mathematicians found that creating your own time to add to phenomena to acquire your concepts of physical reality puts too much power in the hands of mankind. (I suspect there are religious sentiments in this.) [40] Besides, it was complicated and, worse of all, even subjective. The Minkowski system was easier, simpler and more objective;[41] you just have to mention the Minkowski space or ds^2 and move on. It comes with time already embedded in space as part of it---so the whole of space is 'space-time' and every time is also 'space-time'. The

---

40 The Minkowski formula makes time universal again after Einstein, namely as (s=ct); something in general existence mysteriously (harking back to Pythagorean mysticism in mathematics), which can be invoked with the appropriate mathematical symbols; not as something you create in your own local space with the application of points to space, which makes time completely secular. It seems to me that humankind is not ready to accept time as purely secular. Those of us who have already made the necessary psychological adjustments for accepting time as plainly secular are not regarded as normal.

41 Of course, as Einstein has confirmed, it was difficult in mathematics but easy in logic and philosophy; and let me hurry to add that, because of the involvement of time, the whole notion of local time or space-time has philosophical implications, since time is the second most important thing in the world, second only to life itself.

caveat of Professor Eddington was quietly ignored. Soon everybody forgot about this; Eddington and Bertrand Russell were dead; and there was nobody clever enough to notice the discrepancy and question them about it. Of course, that leads to a distortion of relativity, but mathematicians are the arbiters of truth in mathematical physics and they were the ones benefiting from the Minkowski theory, and therefore preserved it. Otherwise it is not true that all space is 'space-time', while all time is also 'space-time'.

Yet it is true that time is always "space-time". You cannot have time without space; not because the space comes with time inside already, but because all time is known and used in units and units only, which can only be had by the application of points to space. There are elements of time in the mind as the internal sense of time, known as the sense of duration. But we have got to link duration to repetitive external cycles to give us usable time in units, as I have explained above. For example, without space we cannot have the year; yet the year is our basic unit of time out of which all other units are derived. Also, the year has to be repeated for time to continue since it is just one unit of time.

This brings a little complication but nothing serious. The reason is because you can only create time, as 'intervals', or as 'time units', as I suppose (because the year is only one unit of time and we derive all other units from the sub-divisions of the year with points or mathematics), with the application of points to space, thus making time a product of space, and therefore 'space-time'. The truth of the matter is that you cannot have time without using points to divide space; it makes time necessarily discrete, being the product of points. Therefore time is always 'space-time, or properly 'space-timed'. But that is all the connection between space and time, except that space is required, again, for displaying time in units as we have in the clock.[42] The clock, any clock, does not give 'flowing time'. It merely reproduces units of time programmed into it. The old mechanical clock based on coiled springs gave the best illustration. The springs are manufactured to release units of time: second, second, second. If one failed to rewind the springs, the clock stopped ticking. The springs provided the clock's energy, but were strictly programmed to reproduce time in specific units only.

After the time is derived in this way, it becomes separate from both the space and the points used in creating it. That is why Einstein made them separate entities in special relativity.[43] For, apart from the condemnation of the Minkowski 4-D geometry which assumes that time

---

42 As discussed above, the poignant question posed by Bertrand Russell comes up again, namely, in the absence of universal time, what really is measured by the clock? (ABC of Relativity, Ch.4.) This is a very serious matter, because if cosmic time is abandoned, there is no time, or any logical explanation for the time we have. The answer, of course, is that the clock does not measure time. It is deliberately programmed to *reproduce* specific units of time: second, second, second, leading to minutes and so forth, to accord with the cycles of the earth, so that about 31,536,000 (or so many) seconds will coincide exactly with the earth's orbit of the sun, called 'one year'. To have more years, we go round the sun again and again and again---hence perpetual time. Units of time in procession give us continuous time. From the Einstein concept of space-time we know that time, since it is produced with points, has got to be wholly discrete.

43 Otherwise he knew of space-time. It was implied in the special relativity concept of local time he adapted from Lorentz, as Bertrand Russell has pointed out. So, for heretics like me, Minkowski was not needed---and Einstein instinctively got it right, initially calling the mathematical interpretations of his theory 'superfluous learnedness' before the mathematicians worked on him to change his mind and go on even to praise Minkowski!

and space constitute one entity by Russell and Eddington, Professor A. N. Whitehead has also pointed out that time and space still pass through nature separately---not as one entity. To add to these, I have humbly suggested the Principle of Mathematical Equivalence above, which can also be used to denounce the Minkowski arbitrary and fictitious formula.

APPENDIX IV:

The misconceptions of time in relativity

It must not be supposed that the problem of time in relativity has been conclusively settled. Relativity is physics. When a problem is solved in physics the solution is always clear, precise in mathematics, and universally applicable; but time in relativity at present is very vague, neither definite nor precise, not least because consideration of time is a philosophical enterprise. The argument is that the original Einstein theory of time can be used to solve the passage and continuity of time. Unfortunately, Herman Minkowski made the question of time in relativity immensely complex and vague, not at all like the original notion proposed by Einstein. Indeed, as a result, the question of time on the whole is destine to keep the philosophers busy for several centuries as their nostrums become footnotes to Einstein instead of Plato. As regards the physicists and cosmologists, as opposed to the philosophers, they believe that the Minkowski theory makes things easy for them; the problem is that it is just not true of the physical world.

Bertrand Russell has said the concept of space-time is perhaps the most important theory Einstein introduced. To me, there is no doubt (no 'perhaps') about it. It is the most revolutionary theory in human history simply because time is second in importance only to life itself---and yet that life cannot even be lived as a well-organised existence without time. That is how momentous time is in human affairs; and Einstein has shown that it is very different from what it has been traditionally assumed to be. Secondly, he insisted that it should be taken as a separate coordinate in the study of phenomena. In the determination of physical reality, because of Einstein time is a co-ordinate in its own right just like the height or length of matter and space are, thus making Man, the observer, part of the observed, since he has to add the time in the 3+1 formula. Those mathematicians who assume, on the Minkowski theory, that time can be incorporated into space with mere mathematics so that we can dispense with the 3+1 formula and the metaphysical role of man in the determination of physical reality, are contradicting Einstein, which is something approaching a hanging offence in science. On the contrary, it is possible that the passage and continuity of time can be conclusively resolved with the original Einstein theory of time as space-time, or local time.

There is obviously fear in some people that time cannot be something we invent by ourselves. Of course, if 'there is no longer a universal time' we have to find out how we get our time.[44] However, nobody is claiming that man invented the whole of time. Rather we have found that we invented how to quantify time by linking the natural sense of time as duration in the mind to external cycles. This sense of duration of anything is obviously connected with the memory mechanism for the retention of images and concepts in the mind.

Let me stress again, and more strongly, that the sense of time is duration in the mind. In his Mathematical Theory of Relativity, Professor Eddington made this absolutely

---

[44] It is not often realised that philosophy is of great importance to science; and, as an example, this is the sort of thing philosophers do behind the scenes to make their suppositions indispensable to science in general; for the philosophers service every branch of science. The phrase 'survival of the fittest' from biology which has passed into general usage in science and linguistics, was coined by a philosopher, not Darwin. All the sciences need philosophical interpretations. In the quotation above from Professor Dingle, he was saying this very strongly in respect of physics; but all the sciences need the same sort of assistance from philosophy, including even mathematics and logic.

clear, as quoted above; and we have got to take that view seriously because the theory of time outlined in this book is based on relativity. Unfortunately the mental sense of duration is not enough. It cannot give time for general use because it is private. The word 'time' is meaningless until it is objectively quantified. We need time in units to apply to the external world---i.e. to mechanise in the clock for general use, so as to be able to tell 'How much time' at a glance---see Appendix I above. This is achieved with external cycles, the most basic of which is the earth-year out of which all other units of time are derived with mathematics. And it is maintained that this is in complete conformity with the Einstein notion of time, and therefore incontrovertible. Above all, it is the only means by which we can logically solve the problems of the passage and continuity of time.

For now, we are told in all earnestness from the discussions above that relativity is not properly understood. This may be so. But actually relativity is only a theoretical system, a suggestion. It is based on the suggestion that physical reality is not homogeneous but fragmented, and therefore subject to different natural laws. This applies to both special and general relativity. Bertrand Russell called it 'a logically deductive system'. In plain language, 'a new philosophy of physical reality' so logically structured that it demands attention, respect and serious study. And these Einstein has certainly achieved. With Einstein alone we are not talking about genius but a godlike intellectual phenomenon never seen on this planet before; he reconstructed the world of physical reality single-handed, that is the reason he is indispensable to both scientists and philosophers.

So Bertrand Russell was absolutely right. Einstein's system is a new logic of physical reality, and it works. But theoretical physics is most unlike the physics we apply in laboratories. Ordinary physics is much more like chemistry; it has consequences. The Nobel Committee was right to award Lord Rutherford the Prize for Chemistry, even though he regarded himself as a physicist, who had rather cheekily claimed that "all of science is either physics or stamp collecting"!

In theoretical physics there are no obvious consequences, so it is difficult to judge the merits of suggestions. Instead, when we get a new theory in advanced physics (rightly or wrongly), three things will happen. I mean, all three will definitely happen in succession, whatever may be the merits of the new proposal. First, we will get interpretations of the basic postulates proposed in such complex settings (or confused formats) from rival theorists that the debate just has to go on; nothing will be settled in the meantime. But because there are no consequences, nobody will get hurt, no machinery will fail to function; avoidable calamities will not occur. The rains will not stop; the sun will not dim.

The most recent example was the eather debacle (or debate). Secondly, we will get accusations and counter accusations of misrepresentations and misunderstandings. The third possibility (because philosophers share with theoretical physic one subject-matter, being the determination of physical reality), will be philosophical interpretations to arrogate the almighty right to shame and discredit some of the factions in the debate, only for philosophers of different schools to turn the tables---and so the debate will be carried on and on. These philosophical discourses are often pretty profound, giving several intelligent interpretations without being able to settle the argument one way or the other. Strangely, that is how we eventually acquire our knowledge of the external world, sometimes referred to as the practice of 'academic freedom'. That is what happened to Plato. And that is what is happening to Einstein as he has come to replace Plato, in fact, to make his basic suggestion redundant, if not completely false, due to the quantum theory.

A careful examination of what has happened to Einstein's theory of time so far betrays elements of all three conditions. First, we are told that 'most definitely' due to Einstein's analysis of 'Order and Simultaneity' there simply is no 'standard or absolute time frame in the universe'. ('Time Frame' or 'Time Reference' means the same thing. It means the logical criterion of validity.) This is generally accepted as true; for it is reinforced by the Lorentz time dilation and local time concepts.

However, it implies that time in the abstract is utterly indefinable, as I have shown above with discussions about the earth-year. The year is indefinable; other time units in use on earth are defined in reference to the year. But the year on its own is logically indefinable. Again, all our time units, down even to the cesium units, are based on the earth-year; they are meaningful only as related to the year; but like the year, on their own (that is in the abstract) none of them can be logically defined. How long, for instance, is a second in logic without reference to something else? The result is that we all have to use the clock, or clocks, based on the earth-year. By this theory of time (as quantified time), the human intellect is built upon the concept of "points and instants". Instants do not exist independently in nature. Only points do; they had to be discovered by man, but they do exist in nature independently—for example, trees constitute points. Before we learnt to put points on paper, we could see that trees dotted the landscape. Thus points constitute the basic instrument of human thought, especially in mathematics from which all science spring. The instants arise from the act of 'consciously' and 'purposely' moving from point to point, confirming the Russellian notion that time is 'relation between points. Hence quantified time is human in origin, except that the internal sense of time (as duration of anything in the mind) must be recognised as making a psychological contribution to the invention of quantified time in that the external cycles used for quantified time (the years, for instance), have to have psychological anchors (meanings) which are the senses of duration of anything in the mind.

Secondly, in the absence of a standard time frame, what does it mean to claim that time intervals in a moving frame are shorter---shorter as against what kind of standard or universal time? What time intervals are they compared with since there is no standard time frame? (Note that you cannot say they are shorter as compared to other clocks outside the moving frame; that will bring in the Einstein theory of frames, as I will discuss presently.)

So we all, in the end, have to resort to using the clock or clocks based on the earth-year. Yet if we use the clocks then it is not correct to claim that time intervals in a moving frame are shorter; they are not naturally or normally (in its proper setting) shorter or longer; they are normal to that frame, or to its natural frame. The moving clock may only seem 'different' as viewed from the outside; but if that is the case then there is no puzzle.[45] The time of the moving frame is not 'our' time; and it is not queer to its natural environment or setting. It is a

---

45 Otherwise it is difficult to see how the behaviour of one clock can affect all time, human physiology and even the material contents of atoms, e.g. muons. If time is defined as the passage of existence in consciousness, how can the behaviour of one clock affect it for all of us? There are still a lot of religious beliefs about time. Time dilation is one of them, so sweet to the religious in science because they can claim that "it is a unique mystery about time predicted by Einstein". In fact, it is not a mystery, let alone predicted by Einstein: he rather solved the little problem with his theory of frames—i.e. the dilated clock belongs to another frame to which it is running normally.

strange phenomenon to those looking in from the outside, in breach of the Einstein theory of frames. In fact, it is irrelevant to anybody but those in the moving vehicle only.

The whole idea of studying other frames from the outside is fraught with difficulties; it can never be an exact science since the standard postulates that make our system work (and make it what it is) might be inapplicable outside our frame, or planet.[46] Speculations into other frames from our frame have been responsible for all the bizarre suppositions about time and space-time from mathematicians and cosmologists in general relativity. I don't think that kind of enterprise is justifiable, especially when it leads to theories that space-time may be infinite in its timelike directions. Space-time cannot be infinite because it is necessarily discrete---the year, for instance, is not infinite. It is only one; all other units of time derived from the year are also discrete and individual. The proper way to think of time as space-time is that its units are in perpetual procession (one year or second following another) to make time seem continuous; as such time can never be infinite.

Nothing illustrates the confusion about time in physics as a result of relativity and how it is misunderstood by scientists than the story of muons. By normal logic they should not last long enough to reach the earth; but they do. With the use of formulaic mathematics and concepts, physicists explain this by saying special relativity provides the answer as follows: the speed of muons is so great that their internal clocks slow down.

---

[46] I think one implication of this is that the laws of physics, or some of them, would differ from ours at least in some parts of the cosmos, if not all over. Einstein was really a very strange genius in physical thought. He introduced the notion of postulates for natural laws in frames. This idea may go very far indeed in the cosmos at large.

Using the theories of time dilation and the so-called twin paradox based on it, it is assumed that as the muons speed and their internal clocks slowed down they age less and thus are able to last long enough to reach the earth. To a logician or philosopher who understands relativity, this is so laughable as to choke him. It is really the best example of the confusion in physics about time in relativity. (1) Time dilation has nothing to do with the muons and how they behave, since time does not dilate internally. Lorentz found that a moving clock would be seen by outsiders as running slowly; but internally those carrying the moving clock would notice absolutely no difference in its performance. Einstein explained this with his theory of frames---the moving clock is in a different frame. There is no logical mechanism for this kind of episode to be able to control time per se. All other clocks would not run slower or faster; and since there is no such thing as absolute time frame, or a standard time, by which all other clocks can be compared, the moving clock's performance has no relevance at all in physics, because its carriers would notice no anomaly; and those outside who notice any anomaly should mind their own business since it is not their time. (2) The idea that muons have internal clocks is based on the Minkowski theory of space-time, where space and time are assumed to constitute one entity; and therefore the reasoning goes that, since the muons occupy space, and all space is space-time, they have their own internal clocks to keep or measure time for them. Again, any logician will describe this as nonsense; for after all, the Minkowski space is known to be fictitious and arbitrary with absolutely no logical validity. Secondly, the very idea is easily disproved thus: we know there are (roughly accurately) specific times on our normal clock for certain events on this planet---let us use Sunrise and Sunset for illustration. If Sunrise is usually 6 am and Sunset is roughly 6 pm as they are in some countries in the Tropics, it is inconceivable that a moving clock can force or influence these times to become 5 am and 5 pm, on the planet just because it is running an hour late---simple.

The reader will have noticed that the name of Lord Bertrand Russell comes up regularly in all discussions of relativity's interpretation. It is inevitable. Russell was highly respected by Einstein, and for very good reasons. He was the greatest philosopher of the time. He was also a great mathematician and logician of genius. A most attractive writer, who won the Nobel Prize for Literature, he wrote about every subject in philosophy, including novels to illustrate moral points. When relativity was announced, he abandoned many of his most cherished ideas as wrong without shame or even mild embarrassment. He was candid and honest in the most adorable way, completely dedicated to the truth no matter how it reflected on his own beliefs. Russell probably had no certain beliefs other than the pursuit of the truth wherever it took him: via science, logic, mathematics or plain common sense, and linguistics. If he was certain that teaching mathematics to people from the cradle could save the world, he would have advocated that as his philosophy.

Concerning relativity specifically, in the later editions of his little book "Problems of Philosophy" he denounced his original philoso-phy as expressed in the book because of Einstein's theories, joking that whoever wrote the original ideas must have been a monkey, but nobody should suppose that the monkey looked, even remotely, like himself! No great philosopher has ever made such a confession; often associated with rulers, they all wrote imperious edicts as if they had discovered the final truth in logic and metaphysics.[47] Indeed, Russell

[47] No surprise, then, that Russell later put them in their deserved places (mostly of dishonour) in his monumental *History of Western Philosophy*. One complaint is that he never even once mentioned the name of Wittgenstein in this great

later called his Fellowship dissertation "somewhat foolish" for the same reason, namely, the geometry used by Einstein had made his discussions of the foundations of geometry completely wrong, and he was happy to admit it and adopt the new Einstein theory. He wrote one of the best interpretations of relativity, still in use, under the title "ABC of Relativity". His book "The Analysis of Matter" can be divided into two. One section is about relativity; the other is mainly about his joint theory with A. N Whitehead to the effect that the world of sense is a construction, not an inference. Yet even this can be traced to relativity, since Einstein made man the observer part of the observed, meaning that man contributes something to the nature of physical reality---i.e. to help with the construction of that reality---and the book was published long after both special and general relativity. It is a moot point. For the Einstein theory was the 3+1 system. The three facets of phenomena are natural; the time is, in Einstein's system, one's own local time. It means one would have to invent his time as a union between the sense of duration and external cycles before having an "objective time for general use

---

book. The reason came from his contemporary, Sir Karl Popper---it was because, "In the long history of philosophy there are many more philosophical arguments of which I feel ashamed than philosophical arguments of which I am proud...Russell saw these things in that light, and so did I..." (From, *Modern British Philosophy*, By Bryan Magee, Secker & Warburg, London, 1971.) In 1959 Russell published his book, *My Philosophical Development*, in which he said he eventually had to reject Wittgenstein because he was talking 'logical mysticism' which was anathema to his basic nature. Of course he was right. Correctly defined, logical mysticism includes religion, mysticism and unscientific gibberish, all dressed up to look like valid logical reasoning with a variety of linguistic trickery. Many aspects of philosophy in Oxford and Cambridge (and elsewhere) remain stuck in this kind of mud ever since.

in one inertial frame" to add to the three natural dimensions of phenomena, to complete the construction of physical reality---or the physical reality relevant to one's frame of reference.

APPENDIX V[48]:

Reply to some critics on the web[49]

In the interest of learning, I am sending you the Appendices of my recent Monograph in which you will find all the answers you seek about my work. However, please understand that I am presenting a rounded philosophical theory about time, how it passes, and how it seems continuous in an attempt to solve the problem of perpetual time without the involvement of God---all of it based on Einstein's notion of time so as to link philosophy to physics by means of time alone. For a very long time I have felt that it has become possible to do so, either by time or by means of the quantum; particularly the quantum because it is the same thing as the light by which we see things as the beginning of human knowledge of the external world.

To be completely rational, epistemology can never ignore the quantum (which is also matter or small pieces of matter), as the very light by which we observe other bundles of quanta as bulky matter---a very intriguing phenomenon, or quandary, in nature. It cannot be ignored in any theory of physical reality, however conceived. To link it to philosophy is to abolish the philosophy that regards physics as 'just another way of looking at the world', and see it instead as the only way, rationally. Even the Platonic theory of Ideas becomes redundant because outside the quanta images cannot exist; and the quanta are seen plainly as light---so we

[48] I am afraid, this Appendix is longer than I would have preferred, simply because my critics have to be treated with respect and answered in some detail. Discussions of ultimate reality cannot (and must not) be treated lightly.

[49] This is the corrected version of the piece posted on the Internet (in a hurry) as my reply to some critics.

see how images are constructed, and by what means. The demise of Idealism is finished off.

Also, because the quantum is time-dependent (as 'energy-second'), I have made one or two comments about it. I don't think it can be the same throughout the universe because the time by which it is known on earth is peculiar to the earth---that is, provided all the universe is not subject to the 4-D geometry of Hermann Minkowski, and therefore a second here is not the same as a second everywhere else in the cosmos, according to the original Einstein theory of time. Thus refuting Minkowski is crucial. As energy-second, the quantum's energy is natural---the time is not. It is our peculiar second; and I have discussed how we make our seconds on this planet at length in the book, suggesting that it could have serious implications for the Theory of Everything.

The nature of time may have a bearing on the Theory of Everything due to the following observation about time. First of all, since Einstein was not a 'professional' philosopher he did not attempt to give the logical grounds why every body in the universe has to have its own time.[50] Unlike mathematics logic is mercilessly dry, acute and uncompromisingly factual; everything must be clearly defined; all conditions and methods clearly spelt out. The Einstein theory of frames is used to justify

---

50 Note that Einstein qualified for the noble title of 'philosopher' for a number of reasons. For instance, the difficulties over the eather or the propagation of light arose simply because the nature of physical reality had been misconceived by both physicists and philosophers---this went all the way to the quantum theory. Only a great philosopher could solve such problems and make the solution part of mathematical physics, and not as an unproven supposition, or suggestion. I will give him the title of 'The greatest Thinker', not the greatest scientists. That honour belongs to Charles Darwin, no matter what religious bigots may say against him.

the claim that time is limited to a frame, but the technical grounds why this is so have never been made clear. That is to say, the conditions in nature that make time limited to a frame have never been clarified. Einstein could not be blamed because he was not writing philosophy.

Let me state the 'necessary logical grounds' why time is limited to a frame in a clear language (without mathematics) for the benefit of the reader or readers: time must be quantified to be useful in science and logic---let us call it 'usable time'. It is meaningless to just mention time as such. Culturally we can only use quantified time, otherwise how could we mechanise it in a clock? Now, to quantify time we have got to employ external repetitive cycles (or regular motions) in association with the internal sense of time, which is the sense of the duration of anything whatsoever, to get our usable periodicities---or time in units; so that the duration is converted to time units, or usable time; until then, time (as duration in the mind) is not usable, and can be sensed only in the passage of existence, motion or silent ageing. These units of time then become unique and applicable only to the body concerned, the body whose cycles yielded them; it is from its cycles that the units were established and so they could not be appropriate to any other body. This is not nit-picking in logic, because on this interpretation of time, all the work of cosmologists in the supposed metric of general relativity is vitiated because they use earth time; yet without this interpretation how we quantify time cannot be explained. I believe that attempts to conceive a Theory of Everything has also suffered from the use of earth time everywhere, when it is obvious that it cannot be applicable every-where: we want to link the quantum to gravity. Yet the quantum cannot be 'a universal unit of energy' because it is time-dependent as energy-second. This must be taken into account, but it is not. Scientists just use the word 'time' and forget about its quantification and unique periodicities.

To get a fair idea of my supposition you will have to read my books about my theory, of which there are more than one. Failing that, these Appendices to my latest work will give you an idea of my philosophy. Please note that those aspects of the Minkowski mathematics you cited have no logical merit. I am questioning his basic premise. I insist that, for time alone, the *i* in his *ict* equation is not tenable; therefore his 's=*ct*...' deductions based on the *ict* equation are wrong.[51] There is no such thing as 'imaginary time'. Mathematicians often forget that mathematical symbols must have causative meanings; but philosophers never forget that. It happens to be one of the obvious differences between philosophy and science. Statistical mechanics in science (as opposed to direct one-to-one causality) overcomes the quirks and deficiencies in the behaviour of phenomena due to the absence of direct one-to-one causality in the nuclear and sub-atomic matter; but that

---

51 At all times it should be realised that, despite the condemnation of some scientists, philosophy is important; believe me, it is very important. Einstein said he was influenced 'very greatly...' by David Hume and E. Mach. To get at what can be considered as the ultimate truth (or the truth for short, if you like), philosophers, unlike mathematicians, have to go to the roots (the logical foundations) of equations; merely repeating the mathematical symbols as written is regarded as shallow, at this level, even an insult. Let me quote part of what Russell wrote about the Minkowski formula---and you cannot say Russell did not understand the mathematics of Minkowski: "...the philosopher cannot but feel dissatisfaction with the apparently arbitrary assumption about intervals..." And again, "...there is great difficulty in suggesting any non-technical meaning for interval; yet such a meaning ought to exist, if interval is as fundamental as it appears to be in the theory of relativity..." (Bertrand Russell, *The Analysis of Matter*, Ch. Xxxviii.)

does not mean the old philosophical causality can be dispense with altogether; for causality still occurs, only statistically. But logically statistical mechanics is also caused. It may not be as direct as throwing a rock to shatter a glass window; it is more like your rock going through intermediaries before reaching the glass window, so that crooked lawyers can disclaim liability; but in logic you're liable for indirectly causing the damage. This is a brief account of the type of causality now envisaged under statistical mechanics. The many mysterious behaviours of sub-atomic and nuclear matter are not without cause; for they occur because those particles exist; if they did not exist, the events associated with them through indirect causality (or statistical mechanics) would not occur. There is so much in physics crying for research in dept which scientists have neglected by relying on the fictitious Minkowski formula. Take the quantum for instance. (As energy-second, the quantum it is time dependent; it materialises periodically in accordance with out time, the units of our time as explained in the section of quantified time. If this time is peculiar to the earth, as I think, then the quantum mathematics cannot be universally applicable; and so the fear that the cosmos contradicts the law of direct causality might be misplaced.[52] I couldn't put it stronger than that. It is sufficient to indicate by this idea that more research is needed, as I suspect that the energy-second which is applicable on this planet might not be universally applicable, and so we cannot rely on the nature of "our quantum" alone to argue that direct causality is cosmically abolished---it may be so from our point of view only, for after all, how important are we. There are stars so immense that millions of our sun will find room in them. Then we must thin of the size of our planet as compared to the sun---and the size of a human being in all that. I suspect that the quantum is

52 Einstein may turn out to be right after all about this matter.

not the end-piece of matter or energy in the universe at large, as opposed to what happens on this minuscule dot of a home for man.)

Back to Minkowski, he cannot hide behind the obvious lack of direct one-to-one causality to try and alter physical reality with his counterintuitive mathematics. Knowing time as it is, where is the imaginary time coming from, and what is it supposed to be like? In other words, what is the meaning of 'imaginary time'? Time, once you think of it, ceases to be any other thing than time in the clock, or quantified time as I have defined it.

The concept of imaginary time (if the reader is not aware), was invented by Hermann Minkowski; that is his *ict* mathematics purporting to equate space to time with counterintuitive mathematics; the i was meant to invoke imaginary time. The idea is arbitrary and therefore logically untenable. Its 'hook' which mathematicians, more theological than physicists, have swallowed 'whole' is completely unacceptable.

Warped Space and Curved Space-Time

This matter deserves a sub-section to itself because all scientists seem to have fallen hopelessly in love with it, quite wrongly, I think. Of course, we all know that the Einstein notion of gravity as caused by warped space has been proved. Then Minkowski came along to claim (merely claim) that space and time constitute one entity, and therefore when space warps, time is also warped. That idea is arbitrary and false because his *ict* equation upon which it is based is logically flawed. It is simply not true that time is intertwined with space and can warped so much so that you could (using the appropriate mathematics) meet your grand parents even before they were born. That is pure mythology, and comes from the Pythagorean 'Trans-migration of souls' long since discredited. If it were true none of us who are not millionaires would still be living on this planet; for wherever our grand parents might be we would join them; anywhere is bound to be better than this world!

More seriously, you will find that I know the Minkowski mathema-tics pretty well to even incorporate it in my corny jokes. At this level every writer is a mathematician of sorts. I even agree that he makes relativity easy to understand from the point of view of mathematics—i.e. by dispensing with the 3+1 formula and still have time inherently in space as a separate co-ordinate. But please (always) remember that Professor Eddington stated that, although useful, we must never forget that the Minkowski theory is fictitious. To me even that is unacceptable. It is not strong enough for me as a condemnation of the Minkowski proposal; useful or not, what is fictitious has no place in physics at all.

Let me digress with a brief mention of something that I know is worrying mathematicians. In discussing time rationally in that peculiar sense of physical reality championed by Ernst Mach (rather than as 'philosophy' or 'mathematics')[53], I have nothing against mathematics. In defence of Ernst Mach, let me say this: it is conceded that there are several aspects of physical reality (or science in general) better described (or

---

53 Mathematics is necessary for creating time in units (the year, for instance, as resulting from a point to a point; and there our time ends, unless we orbit the sun again); but time can never be geometricized because it is not entirely physical. The physical aspects are used merely to quantify it; but there is the inner sense of time, as the sense of duration---and how do you geometrize that? Feeling the sense of duration is as important as the hand of the clock. If the second hand of your clock moved ten paces at a time, you would sense at once that it was not giving you credible duration of time. We take the sense of duration and sub-divide it with external cycles to get time in usable units. That is the end of the relevance of mathematics in the study of time.

written) with mathematics. Some things cannot be understood at all without mathematics. For instance, without mathematics we could not have time as we know it, because we could not state 'how much time' in numerical units; and without that the clock would not exist; civilization would be primitive. This could be a subject in Sci-fi novels---a people without clocks, and therefore condemned to live too close to nature. However, mathematics should not be allowed to dominate the entire field, for the simple reason that you could not demonstrate or write what we know about physical reality by mathematics alone; even if you could do that nobody would understand you.

The suggestion I am making is already graphically illustrated by the life story of the British Nobel Prize winner, P.A.M Dirac, the man who averred that Albert Einstein was the greatest scientist of all time because, "Only scientists like Niels Bohr and Max Planck were qualified to wipe his boots. His theories came out of the blue. They did not follow from what had gone before [and, I would add, yet they work.] [54] Dirac himself was the greatest British physicist since Isaac Newton. However, he was regarded by his peers as a poor communicator, and sometimes incomprehensible. "...as a thinker he was unintelligible except to mathematicians. Even his fellow physicists complained that he worked in a deliberately mystifying private language..." His reply was that, "The quantum world could not be expressed in words or imagined." (Taken from John Carey's review of THE STRANGEST MAN: The hidden Life of Paul Dirac, by Graham Farmelo. Published by Faber, 2008.) Here we have both sides of the argument sufficiently elucidated. The scientific genius wanted to communicate mainly by mathematics. His equally brilliant peers objected that sometimes even they could not make out his meaning. And his response

[54] I am quoting from memory.

was interesting. He claimed that he was not to blame because the quantum theory was necessarily abstruse---and we know he was right. Einstein being 'Einstein', the special theory of relativity was also difficult; and general relativity almost impossible to imagine. So both the genius and his critics are obviously quite right. Quantum theory is abstruse, no doubt about it. According to Niels Bohr "Whoever is not shocked by the quan-tum theory has not understood it." On the other hand, the genius has a duty to make himself understood, otherwise why bother to communicate at all? In parts of his ABC of Relativity, Bertrand Russell warns the reader not to try to visualise what he was describing in general relativity. That is one way of solving the impasse.

This means that what you state with equations (as Bertrand Russell always did) must be rendered in words too, however imperfectly. If you cannot do that your theory will never be able to stand logical scrutiny due to the absence of clarity in definitions. Here is the example I am most fond of: before Minkowski space and time were separate entities. In special relativity "they play distinctive roles", yet it works. How did they come to be one entity after Minkowski? To dwell on his so-called 'counterintuitive mathematics' raises two questions: (a) His mathematics must be faulty, for obviously special relativity works pretty well. (b) Mathematics alone cannot demonstrate the nature of physical reality; thus the Minkowski formula does not accord with the physical reality revealed by special relativity. Hence Mach was right.

I set the Lorentz Transformation of neighbouring co-ordinates (formula) aside so as to avoid the impression that the Minkowski *ict* equation is based on the Lorentz Transformation and therefore acceptable. The common mistake of mathematicians has been to use the Lorentz Transformation of co-ordinates as the basis for deriving the Minkowski formula for equating space to time. This led even the great Einstein to say, "...for to every event there are as many 'neighbouring' events (realised or at least thinkable)"![55] For Goodness sake, equating space to time will never be logically feasible because the time cannot be had without using space, otherwise how do you get it in units? Without space how do you get the year, for instance?[56] Thus the greatest logical mind in science was completely misled by the "Neighbourhood Co-ordinate" formula. Yet there is absolutely no logical method (or route) by which neighbourhood co-ordinates can lead to a natural union between space and time when the time is derived from space in the first place---unless one relies on the Minkowski mythology by supposing that, as Einstein put it, 'it is thinkable'! I do not suppose for one moment that scientists and mathematicians are unaware of the anomaly; surely everybody can see that 'thinkability' is not part of objective reality. I rather think they don't know what to do.

Mathematicians thought the Minkowski formula was a blessing as it makes things easy for them by incorporating time into space; in fact, it has turned out to be a curse, and a very serious one. The 4-D geometry is the ideal solution. Or, rather I should say, it would be the ideal solution if it were true of the natural world.

---

55 From RELATIVITY, op. cit. Part I, Sect. 17. I don't know what Einstein was thinking of when he wrote this. It is an absolute bloomer! However he based it on the transformation of co-ordinates, and so he felt it was right, mathematically.

56 How do you get any unit of time without using space? And once you get your unit of time, using space, how do you put them back together to constitute one entity---with the use of incomprehensible, abstract mathematics? This is a case where the Mach doctrine against the excessive use of abstract mathematics in the study of physical reality is particularly relevant; we have always to remember how greatly that doctrine helped Einstein.

Since it is not true and therefore is untenable in physical reality, it belongs to the realm of fantasy---'Dream Physics', I call it. So they don't know what to do, because they have already incorporated it into physical theory; they can do that, as I have said before, because in theoretical physics no immediate consequences flow from theories[57] ---therefore nobody gets hurt when theories go wrong. Now mathematicians must swallow their pride and agree, as Kurt Gödel argued, that whatever we do certain aspects of mathematics can never be completely objective.[58] This is a clever notion; and it accords with the Platonic simile of the cave. Thus, in relativity, we must revert to the 3+1 formula used by Einstein in special relativity---if it worked there, it would work anywhere else due to the 'Two postulates'. 'Anywhere else' means in any inertial frame subject to the 'Two Postulates' of Albert Einstein. For after all we need physics (or theories of physical activity) to be effective only in an inertial frame, which is the field of special relativity applications, and where we know that the 3+1 formula works absolutely perfectly.

---

57 No such thing as touching a button to produce results.

58 Incidentally, I wish to point out that the Gödel formula, known as the Gödel Universe or Gödel Universes, is vitiated by the Minkowski theory upon which it is based. Every supposition influenced by the Minkowski fiction is bound to be logically flawed. It seems to me a sheer waste of intellectual effort since Eddington, Russell and Professor Whitehead told us that the Minkowski formula is logically untenable. Perhaps scholars have been encouraged to rely on Minkowski because it makes things easy for them, and also because even Einstein accepted it. Let me say that Einstein was coerced. Secondly, he knew that it could not affect relativity because whether time is the same as space or not the second is always the same; and what he wanted *is* that time is incorporated into the study of phenomena as a distinct co-ordinate.

On the other hand, the 4-D geometry is merely 'assumed' to work in general relativity without proof. I actually believe that most of the post-relativity work in general relativity and cosmology has been falsified by the reliance on the Minkowski formula---but mathematicians only have themselves to blame, because Bertrand Russell and Professor Eddington said plainly that the Minkowski theory is fictitious and arbitrary, which meant that, logically, it was only a matter of time before it would be rejected by thinkers.

Yet they did not listen. Objections were regarded as evidence of one's ignorance of counterintuitive mathematics. All the time the proper definition of time was not even attempted. For example, where will the next second come from if the earth stops orbiting the sun? From the obvious answer that it cannot happen (or at least not just yet!), because the earth is gravitationally program-med to always go round the sun, we get the evidence that (a) the time units we use to tell 'how much time' without which our civilization could not survive, are derived from the repetitive motions of the earth; and, as such, are clearly human in origin. That is the proof that we use external cycles in union with the sense of duration to quantify time for cultural use. (b) It also goes to show that the continuity of time is obtained from the succession of time units---the repetition of the year, for instance.

The year is what we sub-divide to get all other units of time on earth. Thus no matter what mathematics are used, it is not possible to equate time (derived from space) to space again! That is contradiction in terms. The whole of post-relativity physics, the real nature of physical reality, and even relativity itself are all distorted by this mistake by mathematicians.

So I rather accept the contrary position taken by the Mathematical Society of Japan, from whose

"Encyclopedic Dictionary of Mathe-matics"[59] I am quoting the following consensus: "Historically, the transformation formula [the equation is stated, but unnecessary here][60] was first obtained by H.A. Lorentz, under the assumption of contraction of a rod in the direction of its movement in order to overcome the difficulties of the ether hypothesis, but his theoretical grounds were not satisfactory. On the other hand, Einstein started with the following two postulates: (i) Special principle of relativity: A physical law should be expressed in the same form in all inertial systems namely, in all coordinate systems that move relative to each other with uniform velocity. (ii) Principle of invariance of the speed of light: The speed of light in a vacuum is the same in all inertial systems and in all directions, irrespective of the motion of the light source. From these assumptions Einstein derived [the Lorentz Transformation] as the transformation formula between inertial systems x = (ct, x, y, z) and x1 = (ct1, x1, y1, z1) that move relative to each other with uniform velocity v along the common x-axis. This was the first step in special relativity, and along this line of thought, Einstein solved successively the problems of the Lorentz-Fitzgerald contraction, the dilation of time as a measure of moving clocks, the aberration of light, the Doppler Effect, and Fresnel's dragging coefficient." The time Einstein used, according to him, was the Lorentz local time, provided, it can be defined as 'time, pure and simple'---meaning it is all

---

59 Published by the Mathematical Society of Japan, The MIT Press, Cambridge, Massachusetts and London, England. Ed. Kiyosi Ito. Vol. II, p. 359 B.

60 The Lorentz Transformation and The General Transformation of Co-ordinates are discussed by Professor Eddington in *The mathematical Theory of Relativity*-- Sections 5 & 15. In any case, the Japanese mathematicians did not think much of the Lorentz Transformation, neither did I, and it is not strictly relevant here either.

there is of time; this qualification is very important. (Appa-rently it can be so defined, for it did not hinder his work until Minkowski intervened.) It means every time is somebody's local time. To overcome that we have learnt to mechanise our time in the clock for general use---but, and this is the crucial point, it is based on the earth's motions. So earth time as derived from the earth's regular motions is all the time we have or can have. This is definitely the view from Einstein; and that was the end of the matter rationally, as Einstein would have it.

Yet, by using concepts like the 'homogeneous Lorentz group', 'time reversal' 'space reflection', 'parity transformation', 'the proper Lorentz group', and so forth, none of which made the original Lorentz formula satisfactory, as the Japanese mathematicians aver, cosmologists, the interpreters of general relativity and pure mathematicians are propounding impossible theories about time and call them Einstenian, but mostly inspired by the Minkowski theory that Russell and Eddington described as arbitrary and fictitious. I hope I make myself clear to avoid any misunderstand-ding.

By the Minkowski theory time travel is said to be 'a scientific possibility'. I am afraid that is not true. All notions of time travel are sheer humbug, because Minkowski was wrong. I have to add that it is quite unacceptable to try to conceal logical errors in thought with mathematics as Minkowski has done. Einstein is not to blame; he was literally coerced by mathematicians to accept the Minkowski formula, but either way relativity is not affected.

There is credible evidence to justify the claim that relativity is not affected whether time is regarded as the same as space, or independent of space. Even I should say that the evidence is not only credible but strictly logical. In the special theory of relativity Einstein made time independent of space. Why didn't he go back to make the time the same things as space in the 4-D geometry after he adopted the Minkowski formula? To me, it is because good old Einstein was no fool. He despised the 'superfluous learnedness' of the mathematicians who were tampering with his theory; yet he needed the support of the scientific public, the majority of whom were coercing him to adopt the Minkowski theory. At first relativity was universally ignored; so he acquiesced; but he was no fool. I believe he knew that either way relativity is not affected. Thus he left special relativity as he originally conceived it---where space is separate from time.

But whether time is the same as space or separate from space, the second is always the same---time is always the same; and Einstein wouldn't miss that point even without his brains. The difference between the two versions of time is philosophical, and it is this: thanks to Einstein we now know that every time is somebody's time; that there is not one (overriding) system of time that covers the whole universe. A second here is not like any other second anywhere else. The Lorentz concept of local time was called 'time, pure and simple' by Einstein, meaning that it is all the time there is, or can be had.

Considering how fundamental time is in human affairs, this is a philosophical concept of a revolutionary kind---there are as many times as there are bodies. Russell put it most succinctly (and I quote from memory): "...there is no longer a universal time which can be applied to any part of the cosmos without ambiguity..." It means our time cannot be applied anywhere else. He's right. So far we know only of the time we have created as our local time to suit the earth's motions. Yet time goes to the roots of our existence. It is closely associated with 'Being' or 'Existence'. There is a natural aspect of time in our conceptions. Professor Eddington called it 'the internal time-sense', being the sense of duration. But we also know that it requires points; therefore it cannot be the same as 'Being'. For it could not have been invented without using points; sentience was required.

Thus we must look for an aspect of life that can be mathematically linked to repetitive external cycles (the years, for example) to yield units of time that accords with physical reality (the most obvious is Day and Night), and can also be mechanised in a clock, which, once achieved, should be seen as man's greatest intellectual (scientific, mathematical and philosophical) creation or invention---and Einstein led the way to the most logical theory ever; that is the reason I maintain that time was Einstein's greatest achievement, for that is how we can logically link physics to philosophy by means of time.

Hence in post relativity physics time must be based on our roots, and I think it is now seen as such. But how? The mechanics must be explained. Well, time is now logically conceived as something whose roots in our minds are based on the sense of duration (or the capacity to experience duration as an internal sense of time, as Professor Eddington put it: that is, of the impressions, images, and so forth, of things enduring, since it takes time to endure or linger), and therefore part of the physical or physiological mechanism for memory---if memory is defined as 'the capacity to repeat'.[61] And that is how we link physics to philosophy. The Minkowski formula contradicts this supposition; as such, his theory would be the greatest achievement of the human mind (linking physics to time via space); but then he based his theory on imaginary time co-ordinates, and therefore it

[61] This goes to the roots of our existence because it is part of the mechanism by which we gain knowledge and remember it---which is the sum of the contents of the human mind; part of that knowledge by which we live is the concept of time, of things lingering. It takes time to linger; added to repetitive external cycles, we get time units to mechanise into clocks.

is regarded as fictitious and arbitrary (by Professor Eddington and Bertrand Russell, no less).

I confess it is (or was) technically difficult but I have somehow managed to show why the Minkowski formula seems to work. This is just an elaboration of the gist of the above paragraph. But for your (and everybody's) benefit it will do no harm to repeat it. The question is whether time is secular and originates on this planet, or it is generally in the universe to be invoked with the appropriate mathematics (and mathematicians seem to prefer the cosmic interpretation that implies the existence of God); secondly, whether i can invoke such a time. It cannot because imaginary time does not exist anywhere except in a dream. However, time (the second) is always the same in our minds whether the time is seen as separate from space or part of 4-D geometry; and that is what Minkowski exploited. The second is always the same in the mind, but how does it get there? Again, I have shown the technical methods in my works as a link between the sense of duration and external cycles---the years, for example---and Professor Eddington said much the same thing (ibid, Ch. 1.8.) Professor Eddington wrote: "The rough measure of duration made by the internal time-sense is of little use for scientific purposes, and physics is accustomed to base time-reckoning on more precise external mechanisms." (ibid, Ch. 1.8---p23, my italics.)

Here is a brief explanation of the idea that Duration x external cycles = time in units, for usable time is in units only, known as 'quantified time': the year is just one unit of time, and all other known units of time are derived from the year with points, thus making them also discrete, including the cesium units, since they have to be related to the second to make sense. To begin with, let us assume there are no clocks: now suppose you see an image on TV, then it goes away after a while. How long it was there is its "duration" in sense or the mind, otherwise known as 'the internal sense of time', or 'the internal time-sense', as Professor Eddington put it. It must be clearly understood that duration implies the passage of time---i.e. during the period (or the life) of an event. But obviously it is not enough; it is not the time you can mechanise in a clock for general application, again as Eddington put it. For cultural purposes[62], something else must be added to the duration, namely, it must be converted to units of time. This is easy to do, for we know how we get the earth-year as a unit of time---round the sun as a cycle. In fact, the year is our basic unit of time, metaphysically. To have more units (or years) we go round the sun again, otherwise there are no naturally occurring time units, or years. Strictly speaking, this is not a theory. The process of creating time units to mechanise into clocks is the metaphysical origin of time as a union between duration and how it is broken down into units---or cycles.

Thus, to convert duration to time units, you will have to use repetitive external cycles like the earth-year.[63] This procedure will be the same for any sentient beings anywhere else in the universe; we can only use repetitive cycles to create time in units. That is the only logical way to obtain time in quantified units---otherwise time is 'silent ageing', 'silent motion', or 'the silent passage of existence', all of which are useless to science and logical thought. Without mathematics logical thought (in abstraction or in any depth) is not feasible[64];

---

62 For any purpose where time is to be cited (as the additional co-ordinate of relativity, for instance), you need to have the time in numerical units, which can only be achieved with external cycles.

63 This is easy to understand. You can even tap your finger, and say, for instance, the duration (or the life of the relevant event) was for so many taps of the finger.

64 We hear of the invention of points being necessary for mathematics. In fact, basically, it was a logical invention

and you need to apply mathematics to duration to get time in numerical units suitable for logical thought. In sleep or coma time will be passing by. But when you come to and want to know the time or how long you've been senseless, you will need mathematics based on some kind of repetitive motions or cycles to be able to have the time in numerical units.

We on earth use the earth-year as sub-divided down to the seconds, or the cesium units, to determine the time "during which" the image was there. The term 'during which' means duration, but it is not enough as time.[65] You will have to relate it to some of the sub-units of the earth-year to get the appropriate time. (You will say the event was 'so-and-so long'; that so-and-so length is obtained elsewhere and applied to the event. Metaphysically there is no other way.[66]) Thus we apply some of the earth's sub-cycles to 'duration' to get the time for it---to get the time 'during which' it was there; and that means converting duration to time by linking it to external cycles. The external cycles themselves do not constitute time, either. They are given durations, or periods of mental lengths (during which they were there), before they can constitute time: a month is longer than a week; and so forth. A second is shorter

---

necessary for mathematics, since mathematics is logical thought in abstraction---e.g. for handling massive volumes and representational reasoning where you cannot see what you are talking about.

65 Because images, impressions, events and so forth, can linger in the mind, they are obviously connected to the mechanism for memory, which is defined in science as "the capacity to repeat".

66 That is why 'time' and 'the application of time' are two distinct operations of the mind, but often they are conflated leading to unnecessary mysteries about time, as discussed earlier.

than an hour. It is by the sense of duration (during the life of an event, an image or an impression) that you can determine which unit of the external cycle to apply to it to get the time, or the number of cycles it was there. The two statements (time and number of cycles) are exactly equivalent.[67] So that you can say it was there for one cycle=one year. Or apply any of the sub-units of the year's cycle to it. Due to our use of clocks we have forgotten that this is how we created and mechanised time for the clock/Calendar system. The metaphysical question is whether there is any other time. Well, the passage of existence is regarded as time, but it's not quantified or usable time. My own opinion is that the method described is the only logical system that any sentient beings in the universe will use to get their quantified time because logical thought must definitely be universally the same everywhere as we have it on earth---for it is a process of reasoning about percepts. Nobody can live rationally in any part of the universe unless he or she adopts reasoning according to percepts, and that is what we call logical thought. It may get more complex and mathematical with increasing volumes (and in non-demonstrative inferences), but ultimately it must be based on percepts. Even Idealism, before it was successfully refuted, was somehow related to percepts: if you cannot see anything the question as to whether it is mental or physical will not arise---simply because the 'it' will not be known. If the Irish philosopher, Bishop Berkeley, ever saw anything (say, the pen he wrote with, the paper he wrote on, the table and chair in his study), then there is no argument. In any case, Berkeley rather confirmed the existence of the quanta hundreds of years ahead, without knowing it, as Bertrand Russell has pointed out in his History of Western Philosophy. The most interesting

---

[67] For example, the numbers of the earth's cycles or orbits round the sun are known as years, or periods of time.

refutation of Idealism, of course, is the quip that a train at a station that can be seen to have wheels cannot be said to lose its wheels when in motion just because the passengers are not seeing them, or looking at them as the Idealist ideology requires.

About myself and reactions to your criticisms, I confess I felt a little sad, not really annoyed but rather sad, that you should mention the Minkowski mathematics to me. At this level it is most unfair to assume that I could be ignorant of the Minkowski mathematics. The truth is that mathematicians have overlooked the caveat of Professor Eddington (in his monumental opus, The Mathematical Theory of Relativity), to insist that Minkowski has changed the nature of time with his formula. To repeat: he said plainly that the Minkowski formula would be ideal for describing phenomena, but, while mathematicians may consider it as useful, they must not forget that it is arbitrary and fictitious (Mathematical Theory of Relativity, Ch.1.1.)

Yet what is the situation now? We find that what Professor Eddington and Bertrand Russell have both described as arbitrary and fictitious is making mathematicians shameful because they allow their works to be guided by it, and out of which such mythologies as time travel become scientifically possible. As always with poor old mankind, people show how clever they are as books about such subjects sell millions and my books are ignored! Even the journals will not publish my papers, and judging from what you say, and seem to believe in so completely, I am not surprised. If Minkowski was right why didn't Einstein go back to amend special relativity?[68] He was coerced to

68 Another question is when cosmologists and mathematicians are going to realise that earth time is not applicable to any other world, frame or metric, outside this planet. All the work they have been doing in general relativity since Einstein is vitiated because Minkowski is wrong and our time is not

use the 4-D geometry in general relativity to make it easier to understand---but he was wrong.[69] As a result of which general relativity is now in a hopeless mess. Yet again, either way the basic postulate of general relativity (the curvature of space for gravity all the way to inferences about black holes) is not affected. Einstein was no fool! Footnotes of his theories will come to replace the footnotes of Plato's theories in philosophy. That is my prophecy. Einstein must have known that, intellectually, he's something like God.

Finally, I have already indicated that we can link physics to philosophy through the relativity concept of time; we can do the same thing, again, through the same Einstein's concept of the quantum, or his "Light Quanta" theory, too. Originally he called it 'a hypothesis'. But it is no longer a hypothesis, not even a theory, as scientists have confirmed it as a fact of nature through QED. So the man I call the only God we know did a lot to justify his divine appellation a hundred times over.

The quantum is seen as light. A single one will be a solitary speck of light of a specific hue; and as small as they are, scientists have invented a machine for counting them one by one (QED is said to be the most

---

applicable to the metric of general relativity---it is a different frame, or world, 'pure and simple' as Einstein would put it.

69 What all those lazy (and also probably religious) scientists want is that time does not have to be added to phenomena in the 3+1 formula as Einstein proposed; they rather prefer time as inherently part of space naturally. But that is based on the Minkowski formula, and the Minkowski formula is logically invalid. It seems we are speaking about two different worlds: one which obeys the laws of logic, and another that lives in the imagination of some scientists because of the Minkowski false theory. Yet since the deaths of Einstein, Eddington and Bertrand Russell, they have incorporated the latter into theoretical physics---it is wrong, but who is to check them?

well-established of scientific theories, and its all concerned with quantum interaction). En masse we call them quanta---or the smallest pieces of matter that can exist. We can link physics to philosophy through the quantum because it is light, and we have always thought we see only by means of light. That thought is rather a misconception. In actual fact, we never see things at all; we see only their images.[70] Physiologically it is quite impossible to perceive anything. Seeing occurs in the brain not on the eye. How, for instance, can we fit a house physically into the brain's tissues? Rather we think we see a thing because its lights consisting of these very small quanta reach the eye with its exact image, and thence to the visual cortex in the brain. So we see only the surface lights of things, not the things themselves. Another philosopher conceived his theory of vision just to account for this very fact. It is precisely what the Irish philosopher, Bishop George Berkeley, stated in his philosophy, although he interpreted it wrongly, as Bertrand Russell put it: "Berkeley advances valid arguments in favour of a certain important conclusion, though not quite in favour of the conclusion that he thinks he is proving. He thinks he is proving that all reality is mental; what he is proving is that we perceive qualities, not things, and that qualities are relative to the percipient." This is a permanent proof that we do not see things; we see only their surface qualities in the form of their exact images as constituted by their light emissions, or radiations, by means of quanta---the smallest bits of matter in existence.

We interpret this as meaning that we see only the smallest parts of matter that, by their nature, can never stand still and always radiating about through their interactions with the electrons of matter. So we see matter, and yet do not see matter. We see only matter's

---

70 Plato conceived his Theory of Ideas to account for this fact.

smallest parts flying about---but they are also matter.[71] The philosophical importance is that pieces of matter do actually fly off things with their exact images; when we capture some of them on the eye we see them, with all their colours because the quanta are naturally coloured. There are intricate technicalities in all this; physicists will not put it so bluntly or crudely; they will have to add a number of fine qualifications. But to philosophers, that is enough to work with. Capturing quanta from things to see the things by them means, as Bishop Berkeley supposed, seeing the surface qualities of things and not the things themselves---the main reason is that all seeing is 'Tele-Vision'. When you are very close to a thing it is difficult to see it properly; if it is too close to the eye you might not be able to see it at all without confusing blurs. Seeing by means of quanta is precisely like photography. The only problem is size.

Size presents a special problem that can only be settled with a dose of speculation as opposed to solid facts. But the speculation is based on the facts given above, and they have been proved more than sufficiently. All that is required is that the inferences based on them should be logically valid.

Now, because it involves the quanta, seeing is obviously like the electronic scanning of (or in) computers.[72] We

---

[71] Because of QED we cannot (at least, I cannot) escape the fact that the quanta are the smallest pieces of matter out of which all other forms of matter have emerged through coalescence. The Scientific Industrial Complex has found another way of spending our taxes, so scientists can go on (who can stop them?) building larger and larger 'Atom Colliders' to split atoms in search of the basic building blocks of matter; but for me, my notions of the origin of all forms of matter begin and end with the quanta, the particles of light, thus making light 'What Is' in logic.

[72] I have to sound the warning that, although the computer can give relevant examples regarding the operations of the

imagine that human vision occurs atom by atom, since the quanta are the smallest sub-atomic particles---they interact with matter through their sub-atomic parts, particularly through the electrons of atoms of matter. This well-known electronic process gives a clue as to how size is handled by the brain. The thousands of pages in a computer file are not spread physically in the computer as we spread papers on a table. Similarly, even a single A-4 paper cannot fit into the brain's 'bloody' tissues. I mean, let's face it, when surgeons cut open the brain they see only bloody tissues; yet these same bloody tissues give us awareness of non-bloody percepts; and we know that is possible only at the sub-atomic level of physical reality. At this point a certain amount of speculation creeps into our thoughts. We believe that seeing the paper (of any colour) involves electronic processing of quanta in the brain, where the bloody tissues our surgeons see are not bloody at all but part of the electronic processing of quanta in things and people's heads. The paper's size is electronically scanned end to end. Since it has its own colour, when it ends, its colour will also cease---and another thing's colour will take over. Let us suppose that the paper is white. At every end of it (all four corners of it) different things' colours will take over. We call the four white corners 'the paper'. The other colours would belong to something else, because the paper has ended, and something else must be there not void. Even void has its own colour to identify its presence. This is how one thing is one thing, and another thing is also another thing in vision. Needless to say, given the nature of human motives, confusion in vision can be induced through this

---

human brain, all such examples are copied from the brain. Let me explain this contradiction in terms. Scientists take suppositions from many sources and test them; some work, others don't; but even those that work are poor imitations (and certainly less complex) of how the brain actually works.

method of visual perception---in magic, for instance. The white sheet of paper will stand out; when it ends, its colour will also end to show that it is no longer there before the eyes. The end of an image is marked by change in colour. As far as the brain is concern, the process of seeing the white piece of paper is like the computer scanning it minutely atom by atom at the quantum level, which is very small indeed.

In this way objects of any size can be visually perceived (electronically scanned), without having the things (of so many different sizes) physically lodged in the brain tissues. Size in vision is determined by volume, shape or form, colour, space and position---all of which can be interpreted as 'different colour patterns'.[73] The light we have to shine on things to see them create billions of quantum points of emission, "scanning"[74] the objects of

---

[73] There is scientific support for this because the best scientific definition of size in vision is called "pattern recognition"---sometimes the phrase is used to represent the actual process of vision itself.

[74] Scanning is not the correct word; but there is no correct word. What is involved is so unique there are no similarities. You have to pity those who write about the quantum with our "crude" traditional means of communication. The nearest description is to say the individual quanta are so small that the total number required to give vision of the wide open sky would fit on to the sharp point of a needle with copious room to spare. And they will persist, making it look like scanning. So long as the thing is there, the quanta radiating from it will persist; therefore vision will occur---the eye is bombarded by the quanta from the thing. Remember that they are trillions x trillions x trillions; they are persisting; they are radiating from all aspects of the thing; thus they will continue to give vision of the thing. The visual process is better describes as the absorption of quanta, the quanta from things---we do not see the images of objects such as is supposed on the old Platonic theory. We absorb their material signals direct from them; if

vision as if with a torch in the dark but from billions of points, thus conveying size in vision. So light does not illuminate objects for us to see them; rather they cause objects to emit lights that carry their own exact images---thus making the Platonic Theory of Idea redundant, as a bonus. Let me repeat this because the Platonists are pretty stubborn: the light source does not illuminate the objects we see. We need the light all right, but it does not illuminate objects. By the miracle of quantum computation and interactions, this is not at all surprising. In fact there is no such thing as illumination in the universe. According to Niels Bohr Light is "Transmission of energy between material bodies at a distance". Thus we should regard the quanta from the light source as interacting with the atoms of the things we see, rather than illuminating them. Emission and absorption of radiation is corpuscular. In the absence of illumination in the universe, Plato was plainly wrong.

The inevitable conclusion is that both science and philosophy (or physics and philosophy) have identified one entity as the ultimate cause of the nature of physical reality, or "What Is", namely: we see the world through the quanta; also in all its physical analysis of the nature of the external world, physics has identified the quanta through QED as the cause of all physical reality in its multitudinous forms. So physics and philosophy are linked in the existence of the quanta, another of Einstein's discoveries. He has therefore done more than we expected of God---even Plato is no longer interesting.

On the other hand, if we are to replace the Platonic Theory of Ideas with the new concept of "Quantum Signals" issuing from things to enable us to perceive them visually, then we have to adopt the corollary,

---

they are big, the visual process is like scanning their surfaces---not quite, but almost like that.

which is "Coded Signals In Perception", meaning that things are given codes in the brain for purposes of memory (and all cognitive processes) and not the things themselves physically, since all things are just too large to fit into the brain as they are seen externally.

This sounds confusing, so I will do my best to explain it. I think electronically coded signals cause vision, and that it is the reason we can dream of objects and of events with the eyes closed. REM seems to indicate selectivity and scanning as in the computer. It is interesting that scientists have found that vivid dreams are particularly associated with REM.[75] The electronic coding may have internal and external aspects. Internally, it is obviously part of the mechanism for memory. Externally, the brain must have created categories of 'perceptive images' (images of classes and groups of objects) over several centuries: the figure of a person, the flight of birds, quadruped motions of animals, shapes and forms of objects, and so forth---a long list of categories, literally infinite. Anything new will get its own category; anything known will have its established category already in place; thereafter associated objects will be added to it. What is on two feet, would be assigned to the category of two-footed creatures, including people; what is flying, to the category of flying objects, and so on. Thus we imagine that when something is apprehended, the brain is able instantly to place it in its appropriate category and recognise it. Internally, the codes of things come into play when the thing is invoked (through the appropriate stimuli); so that however big, it can be seen in the mind's eye, because the coding system is electronic on the quantum level. These are all guesses or speculation, but without them vivid dreams cannot be explained. We are trying to imagine how the brain

[75] By the Sleep Research Centre at Loughborough University, for instance.

works; nobody has any cast-iron proof.[76] But what we have found so far sounds credible. The computer can help, since it works through the system known as 'pattern recognition'; and, as I have said above, in the matter of deciding how size appears in vision, we assume that the quantum signals from things behave as if they are scanning the objects from end to end. So the size of any object will extend to the end of the thing's colour. In this way it is believed that size is determined by shape or form, position and colour---all of which can be abstracted to 'different colour patterns', because even 'the void' has to have its own colour---its colour has to be different from those around it.

There are aspects of the brain physiologically established from scientific medicine, computers, and human behaviours that suggest that the above theory may be close to the truth of how the brain actually functions. Nevertheless, even if the suggestion is proved scientifically, it will take centuries for people to digest these very difficult ideas based on relativity and other ideas of Einstein which are also not yet properly understood even by eminent Professors.

One of the difficulties comes from the fact that, if the quantum is a product of our time, as erg-sec, and the units of our time that give rise to the quantum, are also products of the human mind (through a link between the sense of duration and external cycles), then the nature of human life ought to be re-defined because it is

---

76 However, as his final theory of the human mind, Bertrand Russell said that when a physiologist is examining the brains of a patient, what the physiologist is seeing is in his own brain---see the Chapter, My Present View of the World, in his book, *My Philosophical Development*. This theory of the mind implies a perceptive process that relies on signal codes in visual perception: we imagine that the patient's brain parts invoke the appropriate codes in the physiologist's brain to make him see them---as he would do in dreams with his eyes closed.

not properly conceived either in science or philosophy. For the quantum which is the basis of all matter as far as man is concerned, may not be naturally in existence throughout the universe in the form it appears to us, but rather materialises through our unique way of moulding elements of nature to suit our unique nature; and therefore the 'Copenhagen Interpretation' can be adapted to mean that the strange behaviours of sub-atomic matter may be due to the strange manner in which the human mind is 'generated continually' and interacts (and interferes) with nature, namely, not constituted as a solid matter (or mass) but put together by fleeting and highly perishable impressions at the electronic level---and always growing, changing, interacting and inter-communicating through the neurons, for instance.

It seems nature does not see us as special, or as human beings, but mere neurological robots with no right to claims of superiority or any metaphysical pretensions; perhaps we are more efficient than other animals in nature---well, maybe that implies that we are somewhat superior in a way, but I doubt that that is metaphysically significant. Nobody can have the last word because nobody knows what the origin (or the purpose) of life is.

The importance of what Einstein did, even without knowing it, is this: since the dawn of civilization philosophers have been trying to interpret the world to know what it is made of, or how it is constituted. Out of their inquiries scientists arose to claim that the philosophers have got it all wrong. And they started their own lines of enquiry, the chief part of which is the physical analysis of the external world, known for short as 'physics'. Einstein did not know that the philosophers and scientists have reached a stage in their enquiries where all their theories coincide in the discovery of the quantum---as the most credible candidate of What Is. And the miracle is that man does not even have to infer this idea from any complex theoretical postulates from scientists or philosophers. We see the quantum plainly as light. Of course, it has taken a long time to come to this conclusion, more than one hundred years in fact. Einstein himself did not know it---unaware of the significance of what he had achieved. That is because the work of interpreting his ideas to come to this conclusion was difficult.[77]

The most important thing was the new secular theory of time. Until then time was so mysterious that it had literally become the last hiding place of God after Charles Darwin. I can imagine religious leaders smiling smugly at the increasingly secular theories coming out of physics. But as something that originates from this planet (because there is no longer a universal time), and can be seen as a union between the sense of duration and external cycles, time is liberated from the religions; but it happens to be the most important aspect of life, second only to how the life itself came to be in the universe of inanimate matter, and without which no civilization (to sustain life) could have been possible. With this secular theory of time, added to the theory of quantum, we can now agree that man's science will die with him when the earth ceases to be habitable. Nature is far from uniform; the theories that work for us here on earth will die with us; so let's make the most of our good fortune in the work of Albert Einstein.

---

[77] For nearly half of a century, all my numerous papers to the academic journals were rejected---the editors could not understand what I was saying, and I don't blame them. Even Einstein did not know what he had achieved in this regard. What worries me is that with the Minkowski formula, his lofty theory of time is vitiated; that is why it is necessary to point out that the Minkowski theory was condemned even by Professor Eddington as 'arbitrary and fictitious'.

With the above thoughts in mind, let me hurry to point out that the importance of linking physics to philosophy is to defeat the murderous peddlers of stupid religious myths from all the religions. Since philosophers are strictly logical thinkers, it will put an end to all that buffoonery nonsense from callous religious bigots together with their dung-headed leaders, who claim that science is but one of many ways of studying the world and no superior to any other. Once we eliminate such murderers who promote suicide bombings as passports to Heaven, the only remaining problem, as I see it, is how best to control science rationally to man's utmost benefit all over the planet, so that a scientific forum, organised on the lines of the UN, will come to exist---hopefully---to eliminate mad scientists, who will come, oh yes, they will rise, probably disguised as (religious) scientific prophets!

REFERENCES

Many sources are given in the text. The following are among the major books and papers on the subject with which I am familiar. Readers of my other books will notice that I have retained the same set of references for a number of books. The reason is that I am writing about just one subject---namely time.

ALBERT EINSTEIN (1879-1955)---SPACE TIME, an article in the 1926/27 (13th) edition of the Encyclopaedia Britannica. Also, RELATIVITY, in the same edition.

-------NATURE No. 106, 782, (1921), almost the whole issue was devoted to the confirmation of Einstein's new theory of gravity.

-------The Meaning of Relativity, Princeton University Press, 1966.

-------The Evolution of Physics, (With Leopold Infeld) Cambridge 1838.

-------RELATIVITY, Routledge Classics, London and New York, 2001.

HERMANN MINKOWSKI (1864-1909)----He first mentioned his supposition in a lecture in cologne, known as Raum und Zeit (Space and Time) Cologne 21st September, 1908.

-------Herman Minkowski AdP 47, 927 (1915)

-------Herman Minkowski, Goett. Nachr. 1908 p53. Reprinted in Gesammelte Abhandlungen von Herman Minkowski. Vol. 2, p352. Teubner, Leipzig 1911.

BERTRAND RUSSELL, FRS (1872-1970)----Our Knowledge of the External World, George Allen & Unwin, 1922.

------- Mysticism & Logic, George Allen & Unwin, 1976: a collection of important essays first published in 1917.

-------ABC OF RELATIVITY, George Allen & Unwin, 1958 (recently revised by Professor Felix Pirani---first published in 1925.

-------History of Western Philosophy, George Allen & Unwin, 1946.

-------My Philosophical Development, George Allen & Unwin, 1958.

-------The Analysis of Matter, George Allen & Unwin, 1927.

MORRIS KLINE: Mathematics in Western Culture, Allen & Unwin, London, 1954.

SIR ARTHUR STANLEY EDDINGTON, FRS (1862-1944)

-------The Expanding universe, University of Michigan Press, Ann Arbor, 1933

-------The Combination of Relativity Theory and Quantum Theory, Communication of the Dublin Institute of Advanced Studies, Dublin, 1943.

-------The Mathematical Theory of Relativity, Cambridge, second ed. 1930.

-------The Nature of the Physical World, Ann Arbor, Michigan, 1958.

-------Philosophy of Physical Science, Cambridge, 1949.

-------The Theory of Relativity and its Influence on Scientific Thought, Oxford, 1922.

-------Space, Time and Gravitation, Cambridge, 1920.

SIR JAMES JEANS, FRS: Physics and Philosophy, Cambridge, 1942.

-------The Mysterious Universe, Cambridge, 1930.

-------The New Background of Science, Cambridge, 1933.

PROFESSOR A.N. WHITEHEAD: The Concept of Nature, Ann Arbor, Michigan, 1957.

-------Science and the Modern World, Cambridge, 1922.

-------An Inquiry Concerning the Principle of Natural Knowledge, Cambridge, 1919.

-------Nature and Life, Cambridge, 1934.

-------Process and Reality: An Essay in Cosmology, Cambridge, 1929.

-------Essays in Science and Philosophy, Rider & Co., London, 1948.

-------The Principle of Relativity, Cambridge, 1922.

Professor BANESH HOFFMANN: The strange Story of the Quantum, Dover Pub. Inc. New York, 1959.

Professor STEVEN F. SAVITT (ed.) Times Arrows Today: Recent Physical and Philosophical Work on the Direction of Time, Cambridge, 1995.

CHARLES A. FRITZ: Bertrand Russell's Construction of the External World, Routledge & Kegan Paul, London, 1952.

Professor JEREMY BERNSTEIN: Albert Einstein and the Frontiers of Physics, Oxford, 1996.

Professor RICHARD FEYNMAN: Lectures---The Character of Physical law. MIT Press, 1967. There are several volumes of the Feynman lectures and they are all worthy of serious study.

Abraham Pais, "Subtle is The Lord: The Life and Science of Albert Einstein", Oxford, 1982. Professor Pais has methodically provided details of almost all the original papers relevant to relativity. His list is so exhaustive I don't know of a better one anywhere.

www.ingramcontent.com/pod-product-compliance
Lightning Source LLC
Chambersburg PA
CBHW021958170526
45157CB00003B/1047

* 9 7 8 1 4 4 5 7 2 5 0 1 7 *